AF905108

Geologic Hazards of the Pacific Northwest

Geology Lab Manual

Written and compiled by
Reanna Camp-Witmer, Autumn Christensen,
and Shannon Othus-Gault

with additional contributions from Sheila Alfsen, Andrew Frank,
Bill Orr, and Mariah Tilman

Geologic Hazards of the Pacific Northwest: Geology Lab Manual
ISBN: 978-1-955499-43-9
© 2018, 2019, 2026 by Chemeketa Community College.

All rights reserved. Edition 1 2018. Edition 2 2019. Edition 3 2026
No part of this book may be reproduced or transmitted in any form or by any means, electronic or mechanical, including photocopying, recording, or by any information storage and retrieval system, without permission in writing from the publisher.

Chemeketa Press
Chemeketa Community College
4000 Lancaster Dr NE
Salem, Oregon 97305
collegepress@chemeketa.edu
chemeketapress.org

A full list of credits appears on pages 193-194 and constitutes an extension of this copyright page.

References to website URLs were accurate at the time of writing. Neither the author nor Chemeketa Press is responsible for URLs that have changed or expired since the manuscript was prepared.

Printed in the United States of America.

Land Acknowledgment

Chemeketa Press is located on the land of the Kalapuya, who today are represented by the Confederated Tribes of the Grand Ronde and the Confederated Tribes of the Siletz Indians, whose relationship with this land continues to this day. We offer gratitude for the land itself, for those who have stewarded it for generations, and for the opportunity to study, learn, work, and be in community on this land. We acknowledge that our College's history, like many others, is fundamentally tied to the first colonial developments in the Willamette Valley in Oregon. Finally, we respectfully acknowledge and honor past, present, and future Indigenous students of Chemeketa Community College.

Contents

Name: ______________________________ Lab Time: __________________ Due: ____________________

1

1 Scientific Method: What a Geologist Sees

Purpose

An introduction to geologic hazards and how geologists use evidence to understand Earth processes and history.

Introduction

This intro lab activity shows photos of geologic features (rocks and landforms). You will carefully observe the photos to make observations, inferences, and interpretations. An **observation** (the "what") is a statement based on evidence that one has seen, heard, or noticed (raw facts and data - no explanations).

In geology, observations are objective, measurable, and factual descriptions of rocks, minerals, and landscapes. An **inference** (the "how") is a logical deduction, educated guess, or conclusion reached based on evidence and reasoning (but not the "full story"). In geology, inferences are an attempt to explain how observable features form (the process).

An **interpretation** (the "why") is the bigger picture. It integrates inferences and observations to understand the broader significance. In geology, interpretations place a rock or landscape within the area's overall geological history.

Part 1. Observing Photos of Geologic Figures

Observe the photos of geologic features and make observations, inferences, and interpretations based on what is shown. Use critical thinking but don't agonize over being correct or using formal geological terminology. Review the example provided before proceeding to the questions.

1

Instructions

1. Observe Figure 1.1 and review the observations, inferences, and interpretations given in Table 1.1.

Figure 1.1. Ice-covered trees in Oregon, 2021.

Table 1.1. Observations, Inferences, and Interpretations from Figure 1.1.

Observations	Inferences	Interpretation
» Broken and fallen trees are visible. » Branches are mostly coated in a thick layer of ice. » Fallen or bent branches partially block a house. » The ground is covered in snow and/or ice.	» The branches are carrying a heavy load. » Ice added weight to the branches. » The branches broke after becoming overloaded. » Many other branches besides these shown are likely affected similarly.	» A winter storm passed through the area, causing damage. » The storm event likely also damaged homes, roads, and power lines, disrupting the lives of people in the area.

2. Observe Figure 1.2 and list your observations, inferences, and interpretations in Table 1.2 (using the prompts given).

Figure 1.2. Belknap Crater area, Central Oregon.

Table 1.2. Observations, Inferences, and Interpretations from Figure 1.2

Observations	Inferences	Interpretation
» Describe the size, shape, color, and arrangement of the rocks and the distribution of vegetation.	» Does this landscape appear old or relatively young? Why? » What might explain the lack of soil?	» What geologic process most likely produced this landscape?

1

3. Observe Figure 1.3 and list your observations, inferences, and interpretations in Table 1.3 (using the prompts given).

Figure 1.3. Crooked River area, Central Oregon.

Table 1.3. Observations, Inferences, and Interpretations from Figure 1.3

Observations	Inferences	Interpretation
» Describe the appearances of the two rock types shown. How are they different? » Is the contact between them sharp or gradual?	» Did these rocks form at the same time? » What do their positions suggest about their ages relative to each other?	» Which rock unit formed first? » What sequence of events could have created this outcrop?

4. Observe Figures 1.4a and 1.4b and list your observations, inferences, and interpretations in Table 1.4 (using the prompts given).

Figure 1.4a (top). Crater Lake, Cascades. Figure 1.4b (bottom). Wizard Island in Crater Lake, Cascades.

Table 1.4. Observations, Inferences, and Interpretations from Figure 1.4a and 1.4b

Observations	Inferences	Interpretation
» Describe the shape of the lake basin. » Are the surrounding slopes steep or gentle? » Is the lake located in a valley, on a mountain, or in a depression?	» Did material accumulate here or was material removed? » Does this feature appear to have formed gradually or rapidly? » Did the island form before or after the lake basin?	» Which rock unit formed first? » What process could create such a large depression? » What process likely created the island?

5. Observe Figure 1.5 and list your observations, inferences, and interpretations in Table 1.5 (using the prompts given).

Figure 1.5. Norris Geyser Basin, Yellowstone National Park.

Table 1.5. Observations, Inferences, and Interpretations from Figure 1.5

Observations	Inferences	Interpretation
» Describe the appearance of the landscape shown	» Why might plants be absent near the water? » What does the steam suggest about underground conditions? » The features shown here are responsible for the second-highest number of deaths in the park since its establishment in 1872. What about them do you think is dangerous?	» What is happening beneath the ground to produce these features?

6. Observe Figures 1.6a and 1.6b and list your observations, inferences, and interpretations in Table 1.6 (using the prompts given).

Figure 1.6a (left). The Pinnacles, Crater Lake. Figure 1.6b (right). The Pinnacles (closer up), Crater Lake.

Table 1.6. Observations, Inferences, and Interpretations from Figure 1.6a and Figure 1.6b

Observations	Inferences	Interpretation
» Describe the appearance of the spires (shapes and texture). » Are the spires isolated or connected?	» These spires are on the flank (slope) of Crater Lake (i.e., not within the crater). Do these appear to have been built upward or carved downward? » What might have happened to the material between the pinnacles?	» What process could explain the formation of these spires?

1

7. Observe Figure 1.7 and list your observations, inferences, and interpretations in Table 1.7 (using the prompts given).

Figure 1.7. Big Sur, California.

Table 1.7. Observations, Inferences, and Interpretations from Figure 1.7

Observations	Inferences	Interpretation
» What changes do you notice on the hillside? » Where is loose material located? (circle on image) » Is any infrastructure affected?	» In what direction did the material move? » What force likely caused the movement?	» What geologic hazard is shown? » What factors may have triggered it?

8. Observe Figure 1.8 above and list your observations, inferences, and interpretations in Table 1.8 (using the prompts given).

Figure 1.8. Mt St Helens, Washington State. The photo on the left (older) and the one on the right (younger) were taken just a few months apart.

Table 1.8. Observations, Inferences, and Interpretations from Figure 1.8

Observations	Inferences	Interpretation
» List three differences between the two photographs. » What changes occurred to the summit? » How has the surrounding landscape changed?	» Do the changes appear sudden or gradual? » Was material added, removed, or both?	» What event caused these changes?

9. Observe Figure 1.9 above and list your observations, inferences, and interpretations in Table 1.9 (using the prompts given).

Figure 1.9. Dirt road in Mojave Desert near Ridgecrest, CA.

Table 1.9. Observations, Inferences, and Interpretations from Figure 1.9

Observations	Inferences	Interpretation
» Describe the path of the road. » How far does the offset appear to be? » Is the movement mostly horizontal, vertical, or both?	» Did the road move, or did the ground beneath it move? » What does the offset suggest about motion along the surface?	» What geologic process caused this displacement?

10. Observe Figure 1.10 and list your observations, inferences, and interpretations in Table 1.10 (using the prompts given).

Figure 1.10. Grove of dead trees (Ghost Forest) in Alaska

Table 1.10. Observations, Inferences, and Interpretations from Figure 1.10

Observations	Inferences	Interpretation
» Describe the landscape » Are all trees in this photo in the same condition? » Describe the relationship between the trees and the water.	» Do they appear to have died at the same time or different times? » What might have caused the trees to die?	» What environmental change best explains the evidence? » What geologic process could transform a forest into the landscape shown?

11. Observe Figure 1.11 and list your observations, inferences, and interpretations in Table 1.11 (using the prompts given).

Figure 1.11. Niigata, Japan.

Table 1.11. Observations, Inferences, and Interpretations from Figure 1.11

Observations	Inferences	Interpretation
» Describe the condition of the buildings (as a whole and their orientation relative to the ground). » Do all buildings appear equally affected?	» Did the buildings fail or did the ground beneath them fail? » What might have caused the buildings to have this orientation to the ground?	» What geologic process could explain these observations?

12. Observe the images shown in Figure 1.12a and 1.12b and list your observations, inferences, and interpretations below (using the prompts given).

Figure 1.12a (top). Atatā Island, Tonga 12/28/2021. Figure 1.12b (bottom). Atatā Island, Tonga 1/17/2022.

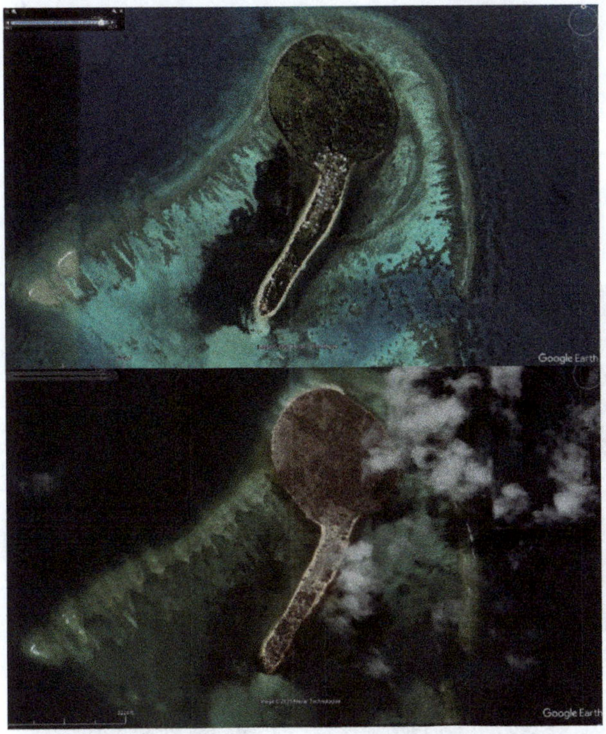

Table 1.12. Observations, Inferences, and Interpretations from Figure 1.12a and Figure 1.12b

Observations	Inferences	Interpretation
» What changes do you notice between the two images? List 3 (and ignore cloud cover in the second image).	» Did these changes likely occur slowly or rapidly?	» What type of geologic process could have caused these changes over only 20 days?

1

1. Which of the hazards shown in this lab do you think poses the greatest risk to people in the Pacific Northwest (Figure 1.13)? Explain your reasoning.

Figure 1.13. The Greater Pacific Northwest.

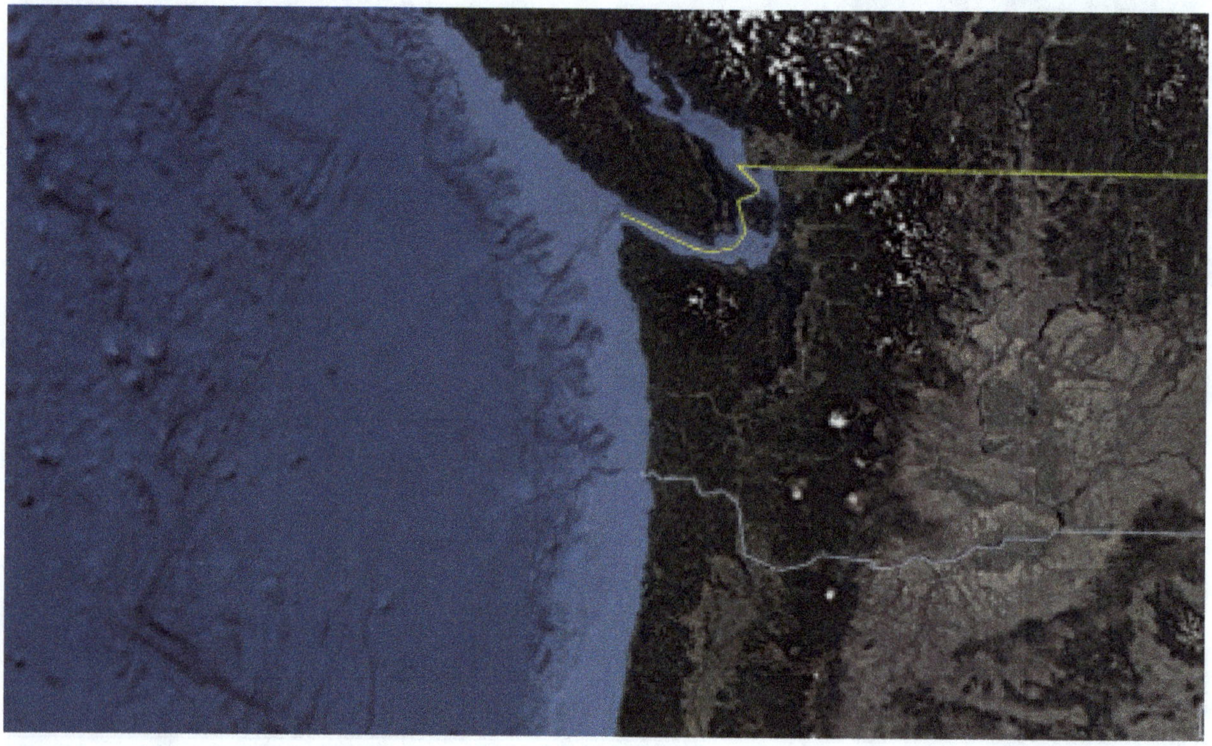

Name: ______________________________ Lab Time: ________________ Due: __________________

2 Plate Tectonics

Purpose

To identify and distinguish between the various plate boundaries and calculate plate velocities.

Materials

- ❑ Calculator
- ❑ Maps
    - » General plate boundary Map
    - » Seafloor Ages Map
    - » Seismology (Earthquakes) Map
    - » Volcanoes Map

Part 1. Magnetic Anomalies

Instructions

1. Use Figure 2.1 and Figure 2.2 showing magnetic anomalies along the Mid-Atlantic Ridge to answer the following questions.

Figure 2.1. Shape of the seafloor, including the mid-ocean ridge.

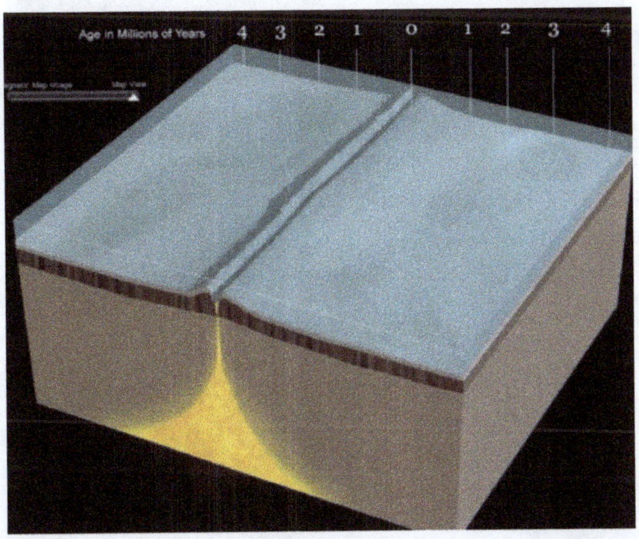

Figure 2.2. Magnetic polarity of the ocean floor over the past 4 million years. Orange represents areas of normal magnetic polarity. White represents areas of reversed magnetic polarity.

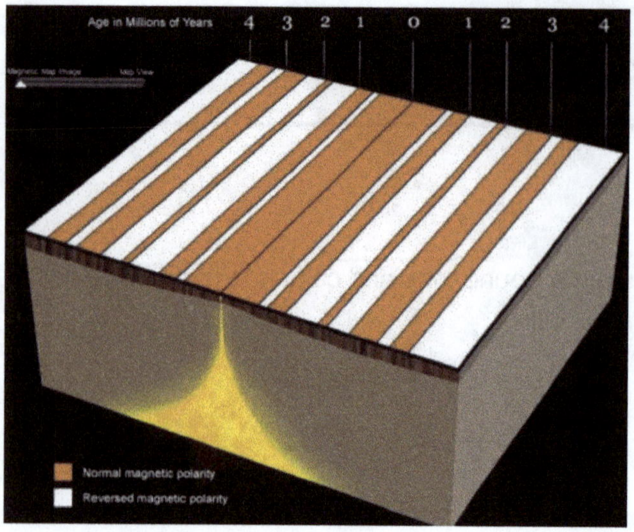

1. How many magnetic reversals (flips) are shown in Figure 2.2 (do not include the current magnetic polarity in your answer)?

2. Approximately how many years ago did the last period of reverse polarity end? In what direction would a compass needle have pointed during this time?

3. Could you expect to find similar magnetic polarity stripes on iron-rich rocks on land? If you could, would those stripes be as useful as the ocean floor in determining past orientations of continents? Explain your answers.

Part 2. Activity Earthquakes, Volcanoes, and Plate Tectonics

Instructions

1. Use the Volcanoes and Earthquakes maps provided. Carefully examine the map keys before you answer the following questions.

4. Match the following landforms on the left-hand side of the page to the associated tectonic plate boundary type on the right-hand side of the page. Some landforms will have more than one plate boundary type. Not all plate boundary types will be used.

_______Mountain Chain	a) Divergent
_______Volcanic Island Arc	b) Oceanic-Oceanic Convergent
_______Volcanic Arc	c) Continental-Oceanic Convergent
_______Mid-Ocean Ridge	d) Continental-Continental Convergent
_______Trench	e) Transform
_______Valley	

5. Which plate boundary type above was not selected for any landform?

6. Explain why the landform indicated in Figure 2.2 is not associated with any major geologic landform.

7. Where are the *deeper earthquakes* (the blue and green dots) located? Give the name of a country or specific region they are primarily associated with. Why do you think these deeper earthquakes are forming in these areas?

8. Generally, in which continents are the majority of *volcanoes* located? Specifically, where do you see a majority of the world's volcanoes?

9. Describe the relationship you observe between the distribution of volcanoes and the occurrence of earthquakes? (There are several possible answers here.)

10. Find a location where volcanoes and earthquakes do not mirror each other (i.e., one but not the other). What processes might explain this difference, and what characteristics make this region distinctive?

Part 3. The Speed of Things

Instructions

1. Use the map in Figure 2.3 and Figure 2.4 to answer the following questions.

Figure 2.3. Political Map of the Mediterranean region.

11. The African Plate is moving at approximately 2.15 cm/yr to the northeast. Using the scale provided, measure the distance (in km) between *Tunis*, *Tunisia*, and *Genoa, Italy*.

12. Using both the rate of plate movement and the distance between these two locations, calculate the time (in years) that it will take for the Mediterranean Sea to close completely. Show your work below.

13. What can we presume will occur when Africa collides with Europe?

Figure 2.4. Map of the Hawaiian Island Chains including ages in millions of years (Ma).

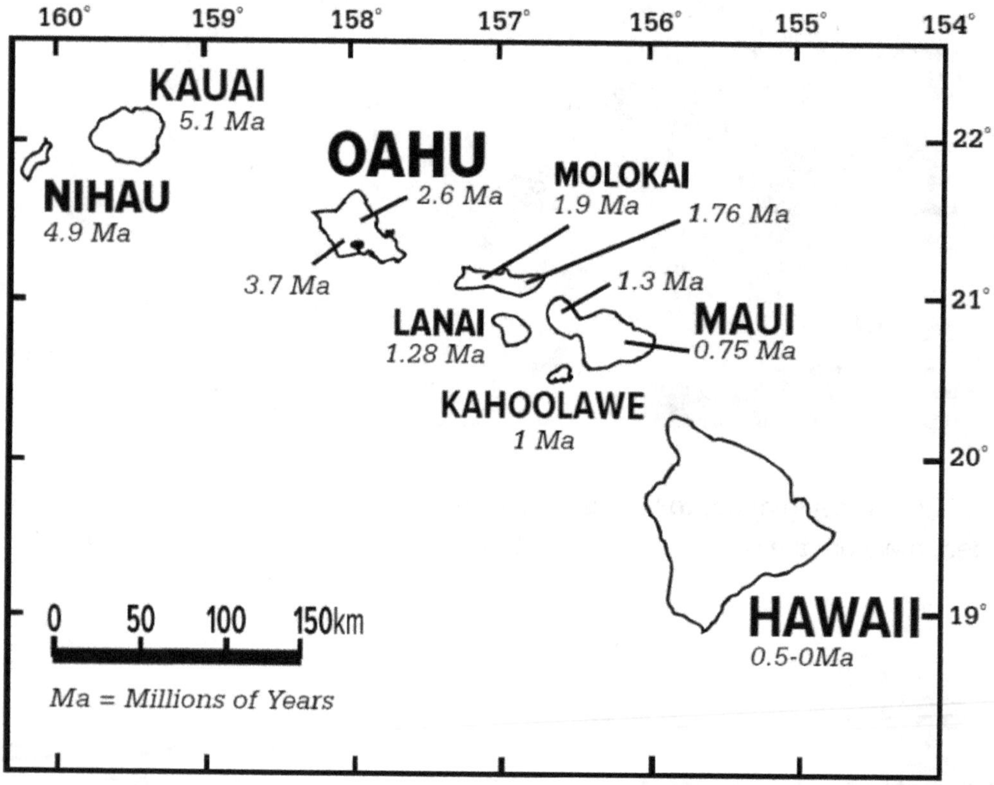

14. Using the ages of the islands of *Hawaii* and *Oahu*, and the distance between the two islands, calculate the speed of plate movement in centimeters per year (cm/yr). Show your work below.

15. Based on the ages of the Hawaiian Islands, what *direction* is the Pacific Plate moving? Draw an arrow on the map above and explain your answer below.

Part 4. Plate Specifics

Instructions

1. Use the map in Figure 2.5 to answer the following question.

Figure 2.5. Tectonic plates in the Pacific Northwest.

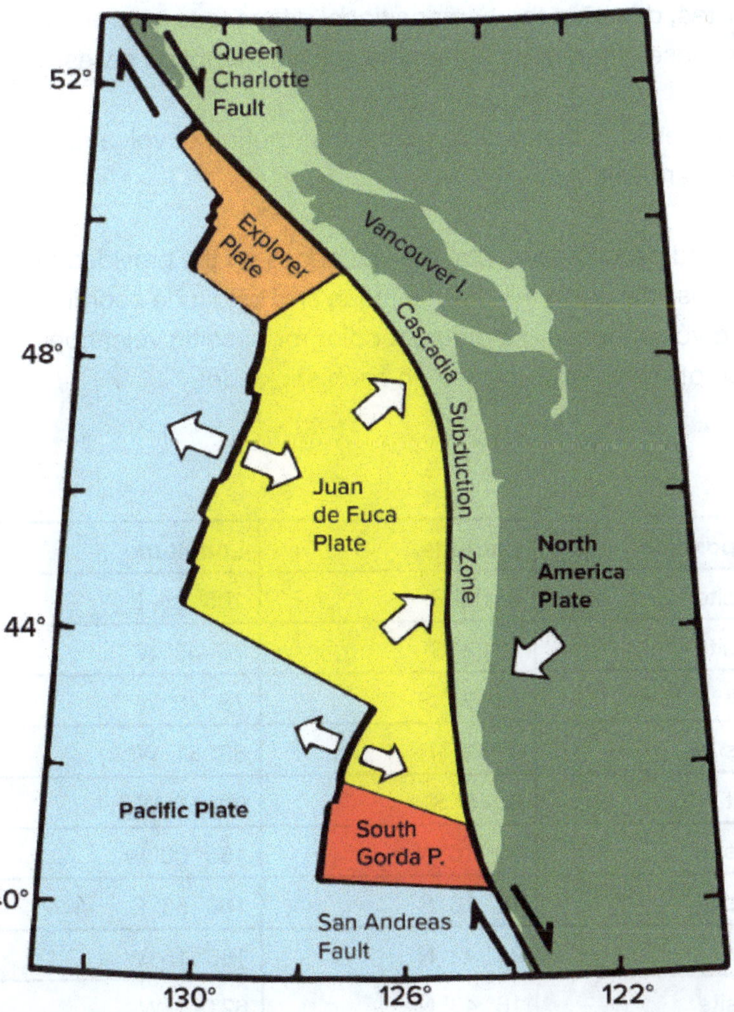

16. Based on the overall southwestern motion of the North American Plate (as shown in Figure 2.5), what is the likely long-term fate of the Juan de Fuca Plate? How would this influence volcanic and seismic activity of the Pacific Northwest over the next ~20 million years?

Part 5. Volcanoes Around the World

Instructions

1. Using your textbook or class notes, describe the *compositional* differences between andesite and basalt. Explain how these differences determine volcanic eruption styles.

The 21 volcanoes listed in Table 2.1 represent a reasonable global distribution of volcanoes and their primary magma type (basaltic and andesitic).

2. Using colored sticky dots or colored pencils/pens, plot the volcanoes on the provided plate boundary map as accurately as possible using the given latitude and longitude coordinates.
3. Use one-color dots for andesitic volcanoes and a different color for basaltic volcanoes.
4. Indicate the number designation of the volcano clearly on each sticky dot.
5. Answer the questions below.

Table 2.1. Types of Igneous Rocks

#	Volcano Name	Composition	Latitude	Longitude
1	Krakatoa	Andesite	6° 06' S	105° 25' E
2	Chimborazo	Andesite	1° 28' S	78° 49' W
3	Juan Fernandez	Basalt	33° 39' S	78° 49' W
4	Irazu	Andesite	9° 59' N	83° 51' W
5	Galapagos	Basalt	0° 45' S	90° 58' W
6	Katmai	Andesite	58° 16' N	155° 00' W
7	Espiritu Santo	Andesite	15° 18' S	166° 53' E
8	Haleakala	Basalt	20° 42' N	156° 15' W
9	Soufrière Hills	Andesite	16° 43' N	62° 11' W

#	Volcano Name	Composition	Latitude	Longitude
10	Ruapehu	Andesite	39° 17' S	175° 34' E
11	Bardarbunga	Basalt	64° 38' N	17° 32' W
12	Santa Maria	Andesite	14° 45' N	91° 33' W
13	Pinatubo	Andesite	15° 08' N	120° 21' E
14	Puyehue	Andesite	45° 35' S	72° 07' W
15	Easter Island	Basalt	27° 7' S	109° 22' W
16	Aconcagua	Andesite	32° 39' S	70° 00' W
17	Eniwetok	Basalt	11° 30' N	162° 20' E
18	St. Helens	Andesite	46° 11' N	122° 12' W
19	Fuji	Andesite	35° 21' N	138° 44' E
20	Kilimanjaro	Andesite	03° 05' S	37° 21' E
21	Nevado del Ruiz	Andesite	04° 54' N	75° 19' W

17. Andesitic volcanoes are located primarily at what type of plate boundary?

Note the locations of basaltic volcanoes in Table 2.1. Some are found on plate boundaries while others are not.

18. Those that are located at plate boundaries are found at what type of boundary?

19. What do we call the type of feature where the "intra" (within) plate basaltic volcanoes are located? Describe this feature.

20. What kind of volcanic material is typical of Cascade volcanoes (one is included in the list)? Based on our current knowledge of igneous composition, what type of eruptions do we expect to see in the PNW?

21. What type of plate boundary does not appear to produce any volcanism?

22. At the boundary of the Indian/Eurasian plate, what feature exists, and why do you suppose there is little to no volcanism here?

Name: ______________________ Lab Time: ______________ Due: ______________

3 Minerals

Purpose

To explore the nature of minerals and familiarize ourselves with some of the most common minerals in Earth's crust, especially the igneous rocks of Oregon.

Materials

- ❑ Beaker or squeeze water bottle
- ❑ Calculator
- ❑ Copper penny
- ❑ Dilute hydrochloric acid
- ❑ Glass plate
- ❑ Graduated cylinder
- ❑ Hand lens
- ❑ Magnet
- ❑ Metric balance
- ❑ Mineral samples
- ❑ Porcelain streak plate

Background

In order to understand the geologic processes that shape the surface of the Earth, you must first have a basic understanding of the nature of the materials that make up Earth's crust. Minerals are the fundamental building blocks of all rocks. When we refer to the "composition" of rocks, we are referring to their mineral makeup, or **mineralogy**. In this and several other lab exercises, you will see how relevant mineral composition is to understanding volcanism and other igneous processes that form the majority of the Pacific Northwest landscape.

Geologically speaking, a **mineral** is a naturally occurring, inorganic solid with an orderly (repetitive) crystalline structure and a definite chemical composition. All minerals can be designated by their chemical formulas and atomic arrangements.

There are over 4,000 named minerals on Earth, and each has a distinctive set of physical properties, such as hardness, cleavage, color, density, luster, fracture, and magnetism. These and other physical properties are commonly used for quick identification of minerals since their internal structure and chemical composition are difficult to measure without elaborate equipment. Most physical means of identification require only the student's skill and some readily available everyday materials. It is important to note that totally pure minerals are rarely found in nature. In general, the more pure the mineral specimen, the more recognizable its properties will be.

The first step in learning to identify minerals is to become acquainted with the properties that, individually or collectively, characterize a mineral species. Carefully follow the sections below, and you will see how these properties are determined using the five mineral samples.

Hardness

Hardness is the resistance of a mineral to abrasion (scratching). Hardness can be determined by scratching a mineral of unknown hardness against a substance of known hardness. The harder substance will scratch or leave a groove in the softer one (i.e., the softer one gets scratched, and the harder one does the scratching).

Hardness is measured on a relative scale using the **Mohs Scale of Hardness**, developed by Friedrich Mohs in 1822. It consists of ten index minerals arranged in order of increasing hardness, where the softest mineral (talc) has a hardness of 1, and the hardest mineral (diamond) has a hardness of 10 (Table 3.1). As a means of comparison, the Mohs hardness values of some common objects are listed in Table 3.2.

Table 3.1. Mohs Hardness Scales for Minerals

Mohs Hardness Scale	
1	Talc
2	Gypsum
3	Calcite
4	Fluorite
5	Apatite
6	Orthoclase
7	Quartz
8	Topaz
9	Corundum
10	Diamond

Table 3.2. Mohs Hardness Scale for Common Items

Mohs Hardness for Common Items	
2.5	Fingernail
3.5	Copper penny
4.5	Wire nail
5.5	Glass or knife blade
6.5	Porcelain streak plate

Testing a Mineral Specimen

Instructions

1. First, test a mineral specimen using the glass plate provided.
2. Find an edge or a corner of the specimen that looks the most like the rest of the mineral (i.e., make sure it is the same color and appearance, not an impurity). It is important that you do not try to scratch the mineral with the glass, as it could chip or break in your hand.
3. Hold the plate firmly on the tabletop while trying to scratch it with your specimen. Very hard substances (H>7) produce deep scratches, but substances of H~6 may produce only faint scratches.
4. If the sample does not scratch the glass (H<5.5), try scratching it with your fingernail. If you cannot scratch it with your fingernail, the sample must have a hardness value of

2.5-5.5. On the other hand, if you can scratch it with your fingernail, the hardness of the mineral is <2.5.

Hint: if it appears that a groove has formed, double-check by brushing it with your finger. Occasionally, the softer substance will "powder off" on the harder substance and give the appearance of a scratch on the harder one.

3

Cleavage and Fracture

The way a mineral breaks provides a lot of insight because it indicates the internal arrangement of atoms within the mineral and the strength of the bonds holding those atoms together. A mineral will tend to break along the plane of its weakest atomic bonds. Cleavage (cleave = to break) describes the tendency of a mineral to break consistently along definite smooth planes (flat surfaces) called cleavage planes. A mineral may exhibit distinct cleavage along one or more cleavage planes, or it may exhibit no cleavage.

Before deciding if a mineral has cleavage, rotate it under a good light source and observe whether there is some position in which the surface of the specimen appears to "light up". Cleavage planes reflect more light than other breakage surfaces, so they appear shinier. Some minerals exhibit "perfect" or "excellent" cleavage with obvious large, shiny, and flat surfaces, sometimes in characteristic shapes. Others exhibit "good" or "poor" cleavage that is less obvious.

Cleavage can be further described by the number of directions cleavage planes and the angles at which they intersect. The number of directions is the number of sets of parallel cleavage planes.

- ❑ Basal cleavage (1 direction) will appear as thin plates or sheets (like the side of a book) when viewed from the "side" (the "top" and "bottom" are parallel, thus it has 1 direction). Minerals in the mica family usually exhibit excellent basal cleavage.
- ❑ Prismatic cleavage (2 directions) is often the most difficult to identify because these minerals rarely break into characteristic shapes. The cleavage planes may intersect at right angles (90°), as in feldspar, or at other angles (e.g., 56° and 124°), as in hornblende. If a mineral with 2 directions of cleavage is a cube, 4 of the sides would be more reflective than the other 2, meaning that the 4 shiny sides are cleavage planes (2 directions) and the other 2 are fracture surfaces.
- ❑ Cubic cleavage (3 directions at 90°) is when cleavage planes meet at 90° (right angles). Often, a mineral with cubic cleavage will easily break into cubes or show numerous right-angle intersections when examined closely. Common examples are galena and halite.
- ❑ Rhombic cleavage (3 directions not at 90°) is similar to cubic, but the intersections of the planes are not at 90°. A mineral with rhombic cleavage might look like a deck of cards where the top has been slid over relative to the bottom. One common example is calcite.
- ❑ Octahedral cleavage (4 directions) looks like two pyramids that share the same square-shaped base. This occurs more often when a sample is pure. Two examples are fluorite and diamond.

If a mineral does not exhibit cleavage, it is said to have fracture. Fracture is typically the result of bonds that are equally or similarly strong in all directions. As a result, breakage surfaces are not flat, but irregular (uneven), conchoidal (smooth curved surfaces or concentric ridges), hackly (jagged edges), or splintery. Additionally, fracture surfaces tend to be much duller than cleavage planes.

3

Although cleavage and fracture describe how a mineral breaks, it is not usually necessary to individually break a mineral specimen because most of the samples you will see have already been broken from larger pieces.

> Note: Do not break a mineral specimen without your instructor's approval!

Specific Gravity

The density of an object is its mass per unit of volume, or in other words, how much it weighs compared to how much space it takes up. It is represented by the formula in Table 3.3, where mass is in grams (g), and volume is in cubic centimeters (cm3).

Table 3.3. The Formula for Density

Mass (g)
Volume (cm^3)

Geologists typically use a related measure, specific gravity, to describe the density or "heft" of a mineral or rock. Specific gravity (SG) is a ratio of the density of a substance (such as a mineral) to the density of water (1.0 g/cm3). This cancels out the units, so essentially it is just an expression of density without units. Most rock-forming minerals have a specific gravity between 2.3 and 3.5, making this property difficult to compare among most samples. Some metallic minerals, however, are noticeably greater [e.g., galena (PbS) is 7.5 and gold (Au) is 17.2].

In Activity 1, you will not be calculating the density or specific gravity of all the mineral samples you will be identifying, but rather comparing this quality among mineral specimens. To do this, it is ideal to select two nearly equal-sized samples and determine which is heavier. Pick up the samples provided one at a time, then try holding one in each hand. You will notice that each seems to have a certain "heft," or specific gravity. In Activity 2 of this lab exercise, we will calculate the density and specific gravity of 5 minerals and/or rocks selected by your instructor.

Color

Color is typically the most obvious property, but it should be documented with caution. Although some minerals have consistent colors, many occur in numerous colors. For example, cinnabar is always red, and malachite is always green, but quartz and fluorite come in a variety of colors. Color variations within the same mineral type can occur due to impurities or defects in the mineral structure. When identifying minerals, it is important to keep this in mind.

Streak

Streak is the color of a mineral in powdered form. This is a useful property because, although the outward color of a mineral can vary, its streak usually does not. Streak is determined by rubbing the specimen on a piece of unglazed porcelain (streak plate), causing part of the mineral to rub off onto the streak plate. This will only work for minerals with hardness less than that of the streak plate (H=6.5), because if the mineral is harder, it will scratch the streak plate instead. Many glassy minerals have a white streak that is difficult to see on a white streak plate. This may only be visible on close inspection. Minerals with a metallic luster typically have a very distinct and easily identifiable streak.

Crystal Form (Habit)

A crystal is a solid that has flat surfaces arranged in a definite geometric shape. The crystal is the outward expression of the mineral's internal atomic arrangement. Crystal form is the common or characteristic shape of a mineral. Since all minerals have orderly atomic structures, all minerals can form perfect crystals if their growth is unrestricted. This is rare, however, because most crystals get crowded together as they grow and their shape becomes less characteristic. Additionally, some minerals have two or more forms. For example, the mineral pyrite is commonly found in either cubes or pyritohedrons (having twelve pentagonal faces).

Knowledge of the various crystal forms can be very useful in the identification of a mineral, but it may also be difficult to differentiate between crystal form and cleavage because cleavage planes may be parallel to crystal faces (e.g., calcite).

Other Properties

The following properties will be useful in identifying only certain minerals.

Reaction to Acid

Certain minerals, such as carbonates (with CO3 in chemical formula), will react with acid by effervescing or "fizzing." When a small amount of dilute acid (typically hydrochloric acid, HCl) is applied to a fresh surface of a carbonate mineral, a chemical reaction occurs, and carbon dioxide (CO2) gas is released as bubbles (Figure 3.5). This is especially obvious in the mineral calcite (CaCO3), which will effervesce vigorously.

Two other carbonate minerals are malachite [(CuOH)2CO3)] and dolomite [CaMg(CO3)2]. These minerals also effervesce, but the reaction is much slower. To magnify the reaction, place a drop of acid on a freshly powdered surface and view under your hand lens. After you have tested a mineral sample with acid, wipe off the liquid with a paper towel. Do not get acid on your skin. Although it is dilute, it can still burn your skin or damage your clothing. If you get acid on your skin, rinse it off with cool water.

Magnetism

Many minerals contain the element Iron (Fe). If a mineral has a high enough iron concentration, it will attract a magnet. To test this, simply hold a hand magnet close to the sample and see if you feel a pull. The mineral magnetite (Fe_3O_4) will strongly attract a magnet, but hematite (Fe_2O_3) will only weakly attract a magnet.

Double Refraction

Double refraction is an optical property present in very few minerals where an image passing through a transparent piece of the mineral appears to be doubled. It occurs when a light beam is split into two perpendicular directions due to the arrangement of the atoms. Calcite ($CaCO_3$) is one example, but note that only a transparent piece will show it.

Striations

Striations are fine, parallel grooves that may be present on the crystal face or cleavage plane of some minerals. Quartz crystals and plagioclase feldspars exhibit striations.

Taste

Some minerals have a distinct taste. For example, halite (NaCl) is chemically identical to table salt (and therefore tastes like it!). Sylvite (KCl), which is often used as a salt substitute for those watching their sodium (Na) intake, has a more bitter taste. Kaolinite, a clay mineral, tastes like aluminum. Do not taste a mineral without asking your instructor first.

Feel

Several minerals have a distinct feel. Talc, for example, has a soapy feel (like a bar of soap) while graphite feels greasy.

Odor

Certain minerals give off a distinct smell. This is especially true if a mineral contains the element sulfur (S). Galena (PbS) and sphalerite (ZnS) both smell strongly of sulfur (like rotten eggs, a match head, or sometimes well water). Kaolinite will give off a distinct "earthy" odor if you breathe on it before sniffing.

Identifying Important Minerals

Using the mineral samples provided at the end of this lab, determine the properties and name of each. Use the mineral tables as a guide.

Table 3.1. Using Mineral Samples to Determine Properties and Names

	Hardness Range	Luster	Breakage (Cleavage pattern or fracture)	Color	Streak	Other Properties	Mineral Name
1							
2							
3							
4							
5							
6							
7							
8							
9							
10							
11							
12							
13							
14							
15							
16							
17							
18							
19							
20							

3

Density and Specific Gravity

Using the designated set of minerals selected by your instructor, determine density (and specific gravity) using water displacement (to determine volume), and mass using the provided metric balance.

Table 3.2. Finding the Density of Minerals and Rocks

Mineral or Rock Name					
Mass of mineral (g)					
Volume of water without mineral (mL)					
Volume of water with mineral (mL)					
Volume of water displaced (mL)					
Volume of water displaced (cm3)					
Density (g/cm3)					
Specific Gravity					

Table 3.3. Percent Error

Mineral/Rock Name (same as above)					
Measured Density Value (g/cm3)					
Accepted Density Value (g/cm3)					
% Error					

1. What were your best (closest) and worst (farthest) measured values from above?

2. What do you think are some sources of error in this experiment?

Metallic Luster						
H < 5½						
Hardness	**Color**	**Streak**	**Luster**	**Cleavage/ Fracture**	**Other Properties**	**Name and Chemical Composition**
1.0-2.0	lead-gray to black	heavy gray	metallic to dull	basal cleavage, easily weathers into fine flakes	low SG (2.1) for mineral with metallic luster; will write on paper; greasy feel	GRAPHITE; C
2.5	bright lead gray	heavy gray	metallic	excellent cubic cleavage	sulfur smell; high SG (7.5)	GALENA; PbS
2.5-3.0	copper (penny)	copper-red	metallic	hackly or irregular fracture	high SG (8.8); often has green discoloration from weathering	COPPER; Cu
1.5-5.5	orange, brown, or yellow	yellow-brown	sub-metallic & earthy varieties	irregular fracture	non-crystalline (amorphous), (see also non-metallic)	LIMONITE; Fe2O3nH2O

Metallic Luster						
H > 5½						
Hardness	**Color**	**Streak**	**Luster**	**Cleavage/ Fracture**	**Other Properties**	**Name and Chemical Composition**
6.0-6.5	brass-yellow	greenish-black	metallic	irregular fracture	"Fool's Gold"; often forms as cubes or pyritohedrons	PYRITE; FeS2
5.5-6.5	steel-gray, dull black, or deep, earthy red, depending on variety	red	metallic (also earthy variety)	irregular fracture	can be weakly magnetic; 3 distinct varieties, relatively high SG (5.3), (see also non-metallic)	HEMATITE; Fe2O3
6.0	dark gray to black	dark gray	metallic to dull black (if weathered)	none	can be strongly magnetic	MAGNETITE; Fe3O4

3

Non Metallic Luster						
H < 2½						
Hardness	**Color**	**Streak**	**Luster**	**Cleavage/ Fracture**	**Other Properties**	**Name and Chemical Composition**
1.0	white, gray, pink, light yellow, or light green	white	pearly, silky	basal cleavage, but weathers easily	very soft; soapy feel	TALC; $Mg_3Si_4O_{10}(OH)_2$
1.5-2.0	clear and colorless to frosty white	white	vitreous to dull	basal cleavage in most varieties (not visible in alabaster)	leaves a frosty appearance when scratched, 3 main varieties (selenite, satin spar, alabaster)	GYPSUM; $CaSO_4{\cdot}2H_2O$
1.5-2.5	bright yellow	white to yellow	usually earthy and opaque	irregular fracture	strong odor	SULFUR; S
2.0-2.5	white, gray-brown	white	vitreous, transparent in thin sections	perfect basal cleavage	part of mica group; peels into thin, flexible sheets or "books"	MUSCOVITE; $KAl_2(AlSi_3O_{10})(OH)_2$
2.0-2.5	white	white	earthy	irregular fracture	easily leaves powder behind on surfaces, can be crumbly, "earthy" smell when damp	KAOLINITE; $Al_2Si_2O_5(OH)_4$
2.0-2.5	dark to pale green	faint green to yellow	pearly or dull	excellent basal cleavage (not visible in compact varieties)	forms short prisms that split into flexible sheets, can also be	CHLORITE; hydrous K-Mg-Al silicate
1.5-5.5 (variable)	orange, brown, or yellow	yellow	sub-metallic & earthy varieties	irregular fracture	non-crystalline (amorphous), (see also *metallic*)	LIMONITE; $Fe_2O_3nH_2O$

Non Metallic Luster						
H 2½ - 5½						
Hardness	**Color**	**Streak**	**Luster**	**Cleavage/ Fracture**	**Other Properties**	**Name and Chemical Composition**
2-6.5	steel-gray, dull black, or deep, earthy red, depending on variety	red	earthy	irregular fracture	can be weakly magnetic; 3 distinct varieties, relatively high SG (5.3), earthy variety can be very soft (see also metallic)	HEMATITE; Fe_2O_3
2.5	colorless to white	white	glassy	perfect cubic cleavage	tastes like salt; soluble in water, transparent to translucent	HALITE; NaCl
2.5-3	dark brown to black	slight greenish	glassy or splendent	perfect basal cleavage	part of mica group; peels into thin, flexible sheets or "books"	BIOTITE; $K(Mg,Fe)_3(AlSi_3O_{10})(OH)_2$
3.0	colorless to white	white	glassy	excellent rhombic cleavage	strongly reacts to dilute acid, transparent varieties may show double refraction	CALCITE; $CaCO_3$
4.0	almost any color (colorless, light green, blue, yellow, purple)	white	glassy	octahedral (rarely perfect in appearance)	exhibits fluorescence, may have slight greasy feel	FLUORITE; CaF_2

Non Metallic Luster						
H > 5½						
Hardness	**Color**	**Streak**	**Luster**	**Cleavage/ Fracture**	**Other Properties**	**Name and Chemical Composition**
5.0-6.0	black (to very dark green)	white	vitreous	two cleavage directions at about 60° & 120°	forms elongated crystals; sometimes bladed; most common amphibole, note that this hardness range spans the hardness of glass (cross listed in table above)	HORNBLENDE; $CaMg_3(SiO_3)_4$, a variety of amphibole
5.0-6.0	very dark green to black	white	vitreous	two cleavage directions at about 90°	blocky, most common pyroxene, note that this hardness range spans the hardness of glass (cross listed in table above)	AUGITE (very complex silicate), a variety of pyroxene
6.0-6.5	usually varying shades of pink ("salmon" pink), sometimes white	white	vitreous to pearly	two directions nearly at 90°	thin veinlets on cleavage planes	ORTHOCLASE; $KAlSi_3O_8$, a variety of potassium feldspar
6.0	dark gray to blue-black	white	high vitreous	two directions nearly at 90°	thin striations on cleavage planes; some varieties will show iridescence	LABRADORITE; (Ca,Na)(Al,Si) $AlSi_2O_8$, a variety of plagioclase feldspar

Non Metallic Luster						
H > 5½						
Hardness	**Color**	**Streak**	**Luster**	**Cleavage/ Fracture**	**Other Properties**	**Name and Chemical Composition**
6.0	white, gray, green	white	vitreous to pearly	two directions nearly at 90°	striations on some cleavage planes	ALBITE; $NaAlSi_3O_8$, a variety of plagioclase feldspar
4.5-6.5	blue to gray-blue	white	vitreous	perfect basal	bladed crystals; hardness varies - 4.5 (long axis), 6.5 (short axis)	KYANITE; Al_2SiO_5
7.0	colorless, white, purple, pink, yellow, brown	white	vitreous	conchoidal or irregular fracture	transparent to opaque, some crystals form hexagonal prisms at one or both ends with fine striations on crystal faces	QUARTZ; SiO_2
7.0	can be white, gray, blue, brown, red	white	vitreous to sub-vitreous to dull	conchoidal fracture	microcrystalline quartz; can form very sharp edges, opaque	CHERT; SiO_2
7.0-7.5	deep red to reddish-brown		Vitreous, resinous, or dull	sub-conchoidal fracture	usually occurs as perfect crystals, numerous varieties	GARNET; complex silicate
9.0	brown to red; sometimes bluish gray		vitreous to dull	irregular fracture, sometimes 1 direction of "false cleavage"	very often hexagonal; SG = 4.0; rubies and sapphires are gem-quality corundum; most samples are opaque and dull	CORUNDUM; Al_2O_3

3

Name: ______________________ Lab Time: ______________ Due: ______________

4 Igneous Rocks

4

Purpose

In this lab exercise, you will learn to identify some of the most common and important igneous rocks. You will also study the various textures and compositions of igneous rocks as well as their origins of formation. Many of these igneous rocks are common in the Pacific Northwest due to the various forms of volcanism (both recent and ancient) in our area.

Materials

- ❑ Glass plate
- ❑ Hand lens
- ❑ Igneous rock samples

Background

Igneous Rocks

Igneous rocks, often called "Fire Rocks" (ignis = fire), form from the cooling (and subsequent crystallization) of molten rock material, called magma. Magma may cool and solidify at great depths (beneath the surface of the Earth to form intrusive (or plutonic) igneous rocks, or it may cool at the surface as a volcanic eruption or lava flow to form extrusive (or volcanic) igneous rocks. Intrusive igneous rocks tend to have relatively coarse-grained textures because magma beneath the surface cools very slowly, thus allowing ample time for growth of large crystals. Conversely, extrusive igneous rocks cool rapidly when exposed to the cooler temperatures of the atmosphere, as lava flows or during violent volcanic eruptions, and therefore exhibit a much finer-grained texture. Mineral composition must also be considered when classifying and identifying an igneous rock. Once texture and composition are determined, the rock name is easily found.

Texture

Because there are a variety of environments in which magmas cool to form igneous rocks, there are several different textures that igneous rocks may exhibit. Igneous texture specifically refers to the size, shape, and arrangement of mineral grains within the rock. Because it is a direct consequence of the rate of magma cooling (as well as a few other physical or chemical conditions), a great deal of information about the environment of formation can be attained by careful study of texture. Below is a discussion of several igneous textures which you will observe in this lab exercise.

Phaneritic (Coarse-grained)

Rocks with a phaneritic texture have crystals that are readily visible with the naked eye. This texture is characteristic of intrusive igneous rocks because it results from large magma bodies that have cooled very slowly (perhaps over hundreds of thousands of years) at great depths (perhaps many kilometers) beneath the surface. Phaneritic textures consist of a mass of intergrown, equidimensional (roughly equal in size) crystals, which are typically 1-5 mm in size. A hand lens is often used to magnify crystals for identification.

Aphanitic (Fine-grained)

Rocks with an aphanitic texture contain microscopic crystals that are generally not visible to the naked eye. This texture usually results from fairly rapid cooling at (or just below) the Earth's surface. Mineral composition for aphanitic rocks is typically determined by a polarizing microscope or inferred from color since properties cannot be determined from such small crystal sizes.

Porphyritic

Rocks with a porphyritic texture have two distinct crystal sizes—larger crystals (phenocrysts) embedded in a finer-grained matrix of substantially smaller crystals (groundmass). Rocks with a porphyritic texture have an interesting cooling history in that they typically represent 2 distinct cooling stages. Because minerals crystallize under a wide range of temperatures, certain minerals can get quite large before other minerals have begun to crystallize. If magma that contains some crystals is suddenly transported to an environment that allows for more rapid cooling (e.g., during a volcanic eruption), the liquid portion of the mixture will crystallize quickly and surround the already formed crystals. A rock with a porphyritic texture is called a porphyry. Porphyritic rocks are further classified based on whether the groundmass is phaneritic (visible crystals) or aphanitic (microscopic crystals). If it has a phaneritic groundmass, it is considered intrusive, whereas if it has an aphanitic groundmass, it is considered extrusive.

Glassy

Rocks with a glassy texture actually lack crystals and are considered amorphous ("without shape"). When molten rock is ejected into the atmosphere, it is quenched so quickly that atoms within the magma do not have time to arrange themselves into an orderly crystalline structure and are instead "frozen in place". This texture is easily recognized as either solid volcanic glass (as in obsidian—discussed in a following section) or as fine shards of intertwined glass (as in pumice—discussed in a following section).

Vesicular

Some fine-grained rocks contain many spherical (or sometimes almond-shaped) voids left behind from trapped gas bubbles. This is often seen near the top of a lava flow because gases in magma rise buoyantly towards the surface. If some gas bubbles do not escape before the lava hardens around them, the resulting rock is said to have a vesicular texture. They can range in size from about 1 mm to 1 cm. Occasionally, these voids, or vesicles, become filled in with another mineral after formation (see amygdaloidal texture below)

Amygdaloidal

An amygdaloidal texture can form in vesicular igneous rocks when minerals have filled in the void spaces. This mineral infilling occurs when hot, mineral-rich groundwater circulates through the empty spaces, slowly precipitating minerals within the vesicles over time. This is considered a secondary process as it typically occurs well after the formation of the igneous rock. The filled-in vesicles are referred to as amygdules and are typically white and composed of quartz, calcite, or zeolite.

Pegmatitic

Rocks with a pegmatitic texture contain abnormally large crystals (> 1 cm) and are referred to as pegmatites. This texture results from very slow cooling and other physical and chemical characteristics of the magma. Most pegmatitic rocks are similar or identical to granite in composition.

Pyroclastic

Rocks with a pyroclastic texture have fragments of mineral grains and ash that have been welded together by the intense heat of a volcanic eruption. Bits of pre-existing volcanic rock may be visible or microscopic.

Composition

Surprisingly few minerals make up most igneous rocks. For coarse-grained rocks, mineral composition is relatively easy to recognize, especially with the aid of a hand lens. For fine-grained rocks, however, mineral composition must be indirectly surmised from the color or shade of the rock. Most minerals found within igneous rocks are silicates. Recall that silicates are considered either "light" or "dark", based on the presence or absence of iron and magnesium. Darker-colored igneous rocks tend to contain a great deal of dark, ferromagnesian minerals, whereas lighter-colored igneous rocks do not. Below is a discussion of the igneous compositions that you will observe in this lab exercise.

Felsic

Light-colored igneous rocks are predominantly composed of light silicates, such as feldspar and quartz. They are referred to as felsic (fel = feldspar, sic = silica) or granitic (from the common crustal rock, granite). Some felsic rocks have 10% or less dark silicates, such as hornblende or biotite mica. Overall, however, most felsic rocks are light in color and have a silica content of 70% or more. Continental crust is predominantly composed of granitic igneous rocks.

Mafic

Dark-colored igneous rocks contain mainly dark, ferromagnesian silicates, such as olivine, pyroxene, hornblende, and biotite mica, as well as calcium-rich plagioclase (such as labradorite), but no quartz. These are referred to as mafic (ma = magnesium, f = ferrium [Fe]), or basaltic (from the common crustal rock, basalt). Mafic rocks tend to be dark in color and denser than felsic rocks and contain less than 50% silica. Oceanic crust is primarily composed of basaltic igneous rock and is also common on volcanic islands and, to a lesser extent, on continents as lava flows.

Intermediate

Intermediate igneous rocks contain roughly equal amounts of light and dark silicates, and are therefore considered somewhat of a "hybrid". The color or shade of intermediate rocks is usually described as medium, or "gray". Rocks of intermediate composition are commonly produced in areas of volcanism as a result of subduction[1]. Volcanoes along the Ring of Fire around the Pacific Ocean tend to be of intermediate composition.

Ultramafic

Ultramafic rocks are almost entirely composed of the ferromagnesian minerals olivine and pyroxene, and therefore are noticeably denser than rocks of other mineral compositions. Ultramafic rocks are relatively rare in the crust, but common in Earth's uppermost mantle. Where ultramafic rocks exist on the surface, they have typically either been pushed upward by tectonic processes called obduction[2] or where unmelted parts of the mantle have been brought to the surface by rapidly rising magma during a volcanic eruption. Perhaps somewhat counter-intuitively, not all ultramafic rocks are dark in color - some are a medium to very light green.

Important Igneous Rocks

Intrusive Igneous Rocks:

Granite

Granite is a very well-known igneous rock because it is commonly used as monumental and decorative stone. It is intrusive, phaneritic, and must contain plagioclase feldspar, potassium feldspar, and quartz. Accessory minerals such as hornblende, muscovite, and biotite are often present in smaller amounts.

Potassium feldspar (such as orthoclase) typically comprises about 80% of the rock. Granite is generally light in color, ranging from light gray to various shades of pink or red. Some varieties are mostly white with some minor amounts of dark silicates. Some granites are further categorized based on color or minor textural differences:

- ❑ Granite Pegmatite is granite with a pegmatitic texture. It tends to have minerals with large, irregularly shaped crystals of gray quartz and feldspar. Some contain large "books" of mica (biotite or muscovite).
- ❑ Granite Porphyry cooled in 2 distinct stages, like other porphyries. Typically, it will have a groundmass of medium-sized grains and phenocrysts of larger crystals.

---

1. Subduction refers to the tectonic process of convergence where an oceanic crustal rock is forced beneath either an oceanic plate or a continental plate.

2. Obduction refers to the tectonic process of convergence where an oceanic crustal plate is thrust onto (above) a continental crustal plate. This is the opposite of subduction is relatively rare since oceanic crust is denser and thinner than continental crust.

Diorite

Diorite is coarse-grained and of intermediate composition. It is usually distinguished by being composed of approximately half light- and half dark-appearing minerals. The majority of its composition is sodium-rich plagioclase feldspar, but includes smaller amounts of amphibole (e.g., hornblende) and biotite, minor amounts of pyroxene (e.g., augite), and sometimes quartz.

Granodiorite

Granodiorite is coarse-grained and of a composition that is between that of granite and diorite (hence, the compound word). It is dominated by plagioclase feldspar (both calcium- and sodium-rich). It has slightly less quartz than granite and more dark, iron-rich minerals (e.g., biotite and hornblende). Setting diorite and granodiorite side by side helps distinguish them by overall color.

Gabbro

Gabbro is a dark, coarse-grained rock of mafic composition. It is predominantly composed of dark calcium-rich plagioclase and pyroxene (e.g., augite), sometimes with small amounts of olivine, hornblende, and magnetite. In some gabbros, the mineral labradorite imparts a distinct play of colors which make the rock valuable for facing buildings and monuments.

Peridotite

Peridotite is dense, coarse-grained, and of ultramafic composition. It is composed entirely of ferromagnesian minerals – predominantly olivine and lesser amounts of pyroxene. It is typically dark to light green. If it is entirely olivine (and overall a light green color), it is classified as dunite.

Extrusive Igneous Rocks:

Rhyolite

Rhyolite is an extrusive, fine-grained igneous rock that is compositionally identical to granite. Because it contains mostly light silicates, it appears as a light pink, buff, gray, or tan color. Often rhyolite will contain small glass fragments, "flecks" of mica, or voids, which are testament to its rapid cooling. Sometimes voids are filled in with other minerals after its formation. Rhyolite can also show flow banding where minerals and glass shards were aligned into bands as the thick lava flowed on the surface.

- ❑ Rhyolite porphyry typically contains light-colored phenocrysts of quartz or plagioclase feldspar.

Andesite

Andesite is an extrusive, fine-grained igneous rock that is compositionally identical to diorite. It ranges in appearance from light to dark gray. Its name is derived from the Andes Mountains of South America, which are almost entirely made of this rock.

- ❑ Andesite porphyry typically contains phenocrysts of plagioclase feldspar or hornblende.

4

Basalt

Basalt is an extrusive, fine-grained igneous rock that is compositionally identical to gabbro. It is typically black, very dark gray in color, or occasionally a dark greenish-black. Basalt makes up the majority of oceanic crust and is the most abundant aphanitic igneous rock on Earth.

- ❑ Basalt porphyry typically contains dark-colored phenocrysts of plagioclase feldspar or olivine.
- ❑ Vesicular basalt is basalt with numerous vesicles (but contains more rock than void space).
- ❑ Amygdaloidal basalt is formed when the vesicular spaces in basalt are filled in with another mineral after its initial formation. These minerals are typically light-colored and spherical or almond-shaped.

Obsidian

Obsidian is compact volcanic glass that lacks crystals. When viscous (thick) molten material cools very rapidly, crystals cannot form before the ions lose mobility and the lava solidifies. Obsidian has a distinctly glassy texture, which makes it easy to recognize. It often contains dark brown or red streaks of hematite. Despite its characteristic dark color, obsidian has a felsic composition and is almost entirely silica. Small amounts of trapped metallic ions give it a dark color. If you can observe obsidian in a fairly thin section (an edge, for example), you can see that it is actually transparent. Because obsidian is predominantly silica, it has a conchoidal fracture and can create very sharp edges (similar to microcrystalline quartz or the sedimentary rock chert!).

Pumice

Pumice is a white to dark gray, very light, low-density, felsic igneous rock. Like obsidian, it cooled rapidly, but instead of being compact volcanic glass, pumice contains numerous, intertwined shards of glass made up of silica. This is apparent when it is observed through a hand lens. When hot gases are extruded from an erupting volcano, lava tends to "froth up" and solidify as the pieces are propelled from a volcanic vent. Pumice is also considered vesicular in texture due to the fact that escaping gas created void spaces.

Scoria

Scoria is a vesicular rock that forms at the uppermost few feet of a basaltic lava flow where rapid cooling causes the gases within it to migrate upward. Before the gas bubbles escape, the lava hardens around it, leaving behind spherical, tabular, or elongated voids. Fresh scoria is black and glassy, but later often turns reddish-brown and dull. Scoria is also sometimes considered glassy, as cooling can be too rapid to form crystals. Scoria can look similar to vesicular basalt, but has a much higher proportion of vesicles and is therefore less dense.

Tuff

Tuff is an extrusive pyroclastic rock that forms during explosive volcanic eruptions with significant ashfall. It is composed of at least 75% lithified volcanic ash, but can also contain a few larger rock fragments (usually pumice, chunks of local rock, and scoria). It is fairly soft, crumbly, and low density. A welded tuff has the same overall composition but is much denser and more compact. This physical difference occurs because when welded tuffs form, the ash is still extremely hot when it is deposited during an eruption. At these high temperatures, the volcanic glass remains soft and "stickier", allowing ash particles and rock fragments to deform and fuse together under their own weight. This welding process produces a stronger, denser rock.

Volcanic Breccia

Like tuff, volcanic breccia is an extrusive pyroclastic rock that forms during an explosive volcanic eruption and contains significant amounts of ash. Volcanic breccia, however, contains a higher proportion of large, typically angular rock fragments (greater than 2 mm in diameter) that are cemented together by a matrix of lithified volcanic ash.

Identifying Rock Samples

For this lab, you will use the igneous chart in Figure 4.1 and the data sheet provided in Table 4.1, and your knowledge of mineral compositions and igneous rock textures in order to identify the numbered rock samples associated with this exercise. The table of common igneous rocks on the following page may be useful. Carefully follow the steps outlined below.

Instructions

1. Select a rock sample (does not necessarily need to be #1)
2. Determine and record the following characteristics in their respective columns on the data sheet:
    a. Texture
    b. Origin (intrusive or extrusive)
    c. Overall color (light, medium, dark)
    d. Minerals (if coarse-grained—be specific—record any minerals you can identify)
    e. Overall chemical composition
3. Based on these characteristics, determine the rock name and record it in the last column.

Figure 4.1. Common igneous rocks and mineral compositions.

Texture ↓ / Composition →	Felsic	Intermediate	Mafic	Ultramafic
Phaneritic	Granite	Diorite	Gabbro	Peridotite
Aphanitic	Rhyolite	Andesite	Basalt	(very rare)
Porphyritic	Rhyolite Porphyry Granite Porphyry	Andesite Porphyry	Basalt Porphyry	(not found)
Glassy	Obsidian Pumice		(scoria)	(not found)
Vesicular	Pumice		Scoria	(not found)

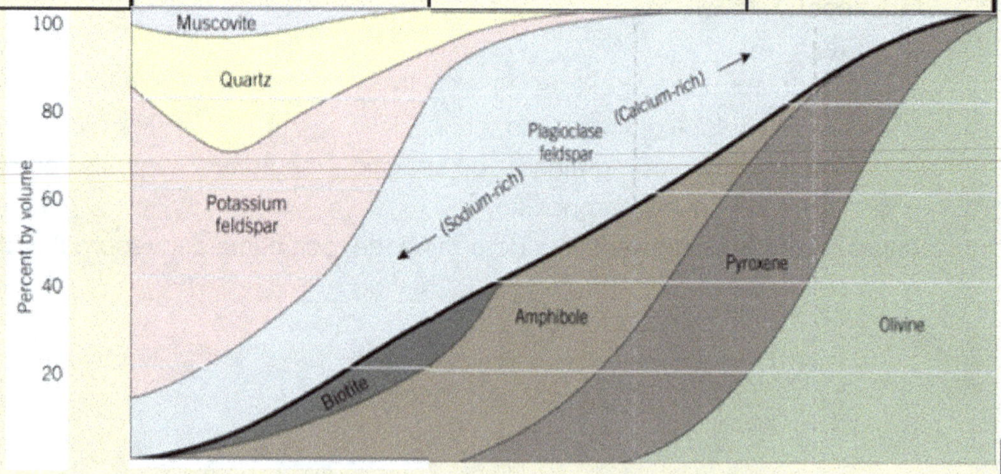

Table 4.1. Data Sheet to Identify the Numbered Rock Samples

	Texture	Origin (Intrusive or Extrusive)	Color (Light, Medium, Dark)	Minerals (Indicate as many as are possible to identify or infer)	Composition (Felsic, Intermediate, Mafic, Ultramafic)	Other Characteristics	Rock Name
1							
2							
3							
4							
5							
6							
7							
8							
9							
10							
11							
12							
13							
14							
15							
16							
17							
18							
19							
20							

4

Name: ______________________ Lab Time: ______________ Due: ______________

5 M&M Magna Chamber

Purpose

In this lab exercise, you will use M&Ms to model the changing mineralogy of a magma chamber in order to understand Bowen's Reaction Series and the process of fractional crystallization.

5

Objectives

- ❑ Explain how fractional crystallization changes magma composition.
- ❑ Relate Bowen's Reaction Series to the order in which minerals crystallize.
- ❑ Calculate changes in magma chemistry as minerals crystallize.
- ❑ Interpret graphs showing compositional evolution of magma.
- ❑ Predict how magma composition influences the type of igneous rock produced.

Key Terms

- ❑ **Atom**: the smallest unit of an element that retains the element's chemical properties. In this activity, each M&M represents one atom of an element.
- ❑ **Element**: a pure substance made of only one type of atom. The elements represented in this activity are silicon (Si), aluminum (Al), iron (Fe), magnesium (Mg), calcium (Ca), and sodium (Na). See also the periodic table of the elements in the Appendices at the back of the manual.
- ❑ **Chemical Formula**: a combination of chemical symbols and subscripts that shows which elements make up a mineral and the number of atoms of each element. For example, Mg_2SiO_4 contains 2 Mg atoms, 1 Si atom, and 4 O atoms.
- ❑ **Mineral**: a naturally occurring, inorganic solid with a specific chemical composition and an ordered crystal structure. Examples in this activity include olivine, pyroxene, plagioclase feldspar, quartz, and magnetite.
- ❑ **Magma**: extremely hot, molten rock material beneath Earth's surface consisting of melt (liquid), dissolved gases, and some crystals.
- ❑ **Crystal**: a solid mineral that forms when atoms arrange themselves into an orderly, repeating pattern as magma cools.
- ❑ **Crystallization**: the process by which atoms in a cooling magma combine to form solid mineral crystals.
- ❑ **Fractional Crystallization**: the process by which newly formed crystals are removed from the magma, causing the composition of the remaining melt to change.
- ❑ **Bowen's Reaction Series**: a model that predicts the order in which minerals crystallize as magma cools.

Materials

- ❑ Calculator
- ❑ Large piece of white paper or poster board
- ❑ M&Ms (or other similar material with at least 6 different colors)

Background

In this activity, each M&M (or alternative material) represents an atom of one chemical element (Si, Al, Fe, Mg, Ca, and Na) in a typical magma chamber[1]. Together, these atoms represent the composition of a fully molten magma chamber. As magma cools, minerals crystallize in a consistent, predictable order (as established by Bowen's Reaction Series). Once crystals form, they are removed from the magma (by fractional crystallization), preventing them from reacting with the remaining melt. Because different minerals contain varying proportions of elements, their removal changes the composition of the remaining magma.

Part 1. Determine the Starting Magma Composition

Instructions

1. Observe Table 5.1, Mineral Compositions. For each mineral listed, use the chemical formula to determine how many atoms of each element (Si, Al, Fe, Mg, Ca, and Na) are present in one formula unit. Record these values in the respective empty boxes.

Example: Mg_2SiO_4 (magnesium olivine[1]):

- ❑ Mg = 2 atoms
- ❑ Si = 1 atom
- ❑ Fe = 0 atoms
- ❑ Al = 0 atoms
- ❑ Ca = 0 atoms
- ❑ Na = 0 atoms

2. Wash your hands before handling M&Ms.
3. Sort the M&Ms by color.
4. Assign each M&M color to the element listed by your instructor.

1. Though oxygen is the most abundant element on Earth (and in magma chambers), it is omitted in this exercise because we would need an impractical amount of M&Ms to represent it.

2. Magnesium-rich end member of olivine group (forsterite).

5. Count the number of M&Ms needed to match the initial magma composition shown in Table 5.1, Mineral Compositions.

Example:

- ❑ Si = 184 M&Ms
- ❑ Al = 70 M&Ms
- ❑ Fe = 39 M&Ms
- ❑ Mg = 40 M&Ms
- ❑ Ca = 34 M&Ms
- ❑ Na = 24 M&Ms

6. Create your magma chamber! Mix all of the M&Ms thoroughly. This mixture represents your starting magma.
7. Place mixed M&Ms at one end of your white sheet of paper or poster board and leave enough empty space at the other end to place minerals that crystallize during the activity.

Table 5.1. Mineral Compositions

	Color	Initial	Mg-rich Olivine	Fe-rich Olivine	Pyroxene	Ca-rich Plagioclase Feldspar	Na-rich Plagioclase Feldspar	Magnetite	Quartz
		Magma	Mg_2SiO_4	Fe_2SiO_4	$CaMgSi_2O_6$	$CaAl_2Si_2O_8$	$NaAlSi_3O_8$	Fe_3O_4	SiO_2
Si		184							
Al		70							
Fe		39							
Mg		40							
Ca		34							
Na		24							
Total		391							

Part 2. Simulate Fractional Crystallization

Important! Before you begin crystallization, remember:

- ❑ The mixed M&Ms represent the magma.
- ❑ The removed M&Ms represent the minerals that have crystallized.
- ❑ Once minerals crystallize, they do not go back into the magma!
- ❑ Oxygen atoms are not represented in this activity. Only count the elements Si, Al, Fe, Mg, Ca, and Na.

Instructions

1. Begin with Crystallization Step 1 using Table 5.2, Mineral Crystallization Steps.
2. For each mineral listed in the step:
    a. Determine how many atoms of each element are contained in one formula unit of that mineral.
    b. Multiply the number of atoms in the formula by the number of minerals that crystallize during that step.
    c. Remove the corresponding number of colored M&Ms from the magma chamber.
    d. Place the removed M&Ms in a separate pile labeled “Crystallized Minerals.” Do NOT return these M&Ms to the magma chamber during the remainder of the activity.
    e. Simulate magma by removing appropriate numbers of M&Ms at each crystallization step. Minerals (M&Ms) that are crystallized in each step are moved to the other end of the magma chamber.

Table 5.2. Mineral Crystallization Steps

Mineral	Step 1	Step 2	Step 3	Step 4	Step 5	Step 6	Step 7	Step 8	Step 9	Step 10
Mg-rich Olivine	2	3	4	3	2					
Fe-rich Olivine			1	1	1	1				
Pyroxene				1	1	4	2	2	1	1
Ca-rich Plagioclase		1	2	3	3	4	3	3	1	1
Na-rich Plagioclase						1	3	6	5	7
Quartz									1	4
Magnetite					1	3	2	2	1	1
Total	2	4	7	8	8	13	10	13	9	14

Important!

- ❑ Step 1 requires crystallizing 2 Mg-rich olivines.
- ❑ Mg-olivine has the formula Mg_2SiO_4, meaning that each Mg-olivine crystal has 2 Mg atoms and 1 Si atom.
- ❑ Therefore, since you are crystallizing 2 Mg-olivine crystals, you will remove 4 Mg-colored M&Ms and 2 Si-colored M&Ms.
- ❑ Place these M&Ms in the Crystallized Minerals pile.

3. After completing Step 1, carefully count the remaining M&Ms of each color still present in the magma chamber.
4. Record the remaining number of each element in Table 5.3, M&Ms Remaining, in the Step 1 column.
5. Repeat 6-9 for each remaining crystallization step until all crystallization steps have been completed.

Table 5.3. M&Ms Remaining

Element	Initial	Step 1	Step 2	Step 3	Step 4	Step 5	Step 6	Step 7	Step 8	Step 9	Step 10
Si	184										
Al	70										
Fe	39										
Mg	40										
Ca	34										
Na	24										
Total	391										

5

Part 3. Calculate Magma Composition

Tips for Calculations:

- ❑ Count only the M&Ms remaining in the magma pile.
- ❑ Recalculate the total number of M&Ms remaining after each crystallization step.
- ❑ Calculate percentages using the current total, not the original total of 391.
- ❑ Check that the percentages for each step add up to approximately 100% (small differences may occur due to rounding).

Instructions

1. After completing each crystallization step, calculate the total number of M&Ms remaining in the magma chamber.
2. Calculate the percentage of each element remaining using the following equation:

   % of element = number of that element remaining | total M&Ms remaining | x 100

3. Record each percentage in Table 5.4, Magma Composition.
4. Calculate the fraction of the original magma that remains after each crystallization step using the equation below (where 391 represents the total number of M&Ms in the starting magma).

   fraction remaining = total M&Ms remaining | 391

5. Record the fraction remaining in the final row of Table 5.4, Magma Composition.

Table 5.4. Magma Composition

Element	Initial	Step 1	Step 2	Step 3	Step 4	Step 5	Step 6	Step 7	Step 8	Step 9	Step 10
Si	47.1										
Al	17.9										
Fe	10.0										
Mg	10.2										
Ca	8.7										
Na	6.1										
Total	100.00										

Questions

Bowen's Reaction Series and Fractional Crystallization

1. Describe how the minerals crystallizing from the magma change throughout the ten crystallization steps. In your answer,

2. Identify which minerals crystallize first, middle, and last.

3. Explain how your observations relate to Bowen's Reaction Series.

Igneous Rock Composition

4. Use the minerals that crystallized during the 2nd, 6th, and 10th crystallization steps to determine the most likely phaneritic (coarse-grained) igneous rock produced during each stage in Table 5.5.

Table 5.5. Recording the Igneous Rock Stages

Step	Rock Name	Evidence (Minerals)
2		
6		
10		

Graph Interpretation

Examine your graph of magma composition and answer the following questions:

5. Which element shows the largest decrease during crystallization?

6. Which element shows the largest increase during crystallization?

7. Explain why these trends occur based on crystallizing from magma.

8. Which element shows the smallest overall change in percentage during crystallization? Use evidence from your graph to support your answer.

5

Magma Classification

Magmas are broadly classified by their silica composition according to the following scheme:

9. Using the classification scheme, classify the magma at the following steps in Table 5.6.

Table 5.6. Classifying Magma

Step	Rock Name
Initial	
2	
6	
10	

10. Describe how the magma evolves throughout crystallization. What overall trend do you see?

Silica Enrichment

11. Silicate minerals are removed throughout the crystallization, yet the silica content of the remaining magma increases. Explain why this occurs.

The Model

12. Which aspects of this model magma chamber are realistic? Which are not? Suggest some ways to make the model more realistic.

5

Eruption Style

13. Based on the composition of the initial and final magma, predict which magma would be more likely to produce an explosive volcanic eruption. (Hint: incorporate the concepts of viscosity and gas-trapping.)

5

Name: ____________________ Lab Time: ____________ Due: ____________

6 Understanding Viscosity

Purpose

In this lab exercise, you will explore the viscosities of several fluids and observe how temperature affects viscosity. You will then apply this understanding to investigate the relationship between the viscosity of magma, its liquid and gas content, and the explosive potential of volcanic eruptions.

Objectives

- ❑ Measure and compare the viscosity of several common fluids.
- ❑ Explain how temperature influences fluid viscosity.
- ❑ Describe how viscosity affects the movement and shape of gas bubbles.
- ❑ Interpret how magma viscosity influences volcanic eruption style.
- ❑ Apply observations from fluid experiments to volcanic hazards in the Pacific Northwest.

Key Terms

- ❑ **Viscosity**: the resistance to flow or deformation of a material (typically a fluid).
- ❑ **Poise**: the unit of viscosity, signified as η ($1\ \eta = 1$ g/cm·s).
- ❑ **Effusive eruption**: a volcanic eruption characterized primarily by lava flows. These are often described as "gentle" and are associated with fewer volcanic hazards.
- ❑ **Explosive eruption**: a volcanic eruption characterized by explosive blasts of pyroclastic materials. These are often described as "violent" and are associated with significant volcanic hazards.
- ❑ **Silica content**: the amount of silicon and oxygen ($SiO2$) dissolved in magma.
- ❑ **Boyle's Law**: describes the inversely proportional relationship between pressure and volume of a gas (i.e., when pressure is reduced, volume is increased and vice versa).

Materials

- ❑ Aluminum foil
- ❑ Beakers
- ❑ Bin
- ❑ Calculator
- ❑ Dropper or syringe
- ❑ Graduated cylinder
- ❑ Hot plate
- ❑ Ice
- ❑ Liquids (common examples below, others may be provided)
  - » Water
  - » Corn syrup
  - » Molasses
  - » Vegetable oil
  - » Glycerin
- ❑ Protractor
- ❑ Ramp (e.g., a clipboard, baking sheet)
- ❑ Ruler
- ❑ Stirrer
- ❑ Stopwatch
- ❑ Thermometer
- ❑ Towels for cleanup

Background

Viscosity is an important property of fluids. It refers to the resistance of a material (usually a liquid or a gas to flow or deformation). Highly viscous liquids are often thought of as "thick" or "sticky," while low-viscosity fluids are "thin" or more "runny." The chemical composition and internal molecular structure of a substance are the most important influences on viscosity, but temperature also plays a significant role.

Although magma is much more complex than the fluids used in this activity, these experiments model many of the same physical processes that govern magma flow and volcanic gas escape. The style of volcanic eruption is predominantly influenced by the viscosity and gas content of magma. Magma viscosity is primarily determined by its temperature and silica content. In general, cooler, silica-rich magma is more viscous and therefore more likely to produce explosive eruptions. Hotter, silica-poor magma is less viscous and is more likely to produce effusive eruptions.

6

Before you begin:
Before collecting data, observe the fluids available to use for this activity and discuss the following predictions with your lab group:

- ❑ Which fluid do you predict will have the highest viscosity?
- ❑ How do you think increasing temperature will affect viscosity?
- ❑ Do you think changing the ramp angle will change the actual viscosity of a fluid? Why or why not?

Part 1. Investigating Viscosity

Here we will measure and compare the influence of substance type (various provided fluids), temperature, and angle of flow on viscosity as measured by flow rates down a ramp. The equation for calculating a fluid's viscosity is shown below.

$$\eta = \frac{g \cdot h^2 \cdot \rho \cdot \sin(A)}{3V}$$

Where:

- ❑ "η" is the coefficient of viscosity (measured in poise, where 1 poise is 1 g/cm·s).
- ❑ "g" is the acceleration due to gravity (980 cm/s^2), a constant.
- ❑ "h" is the depth of the liquid at its foot (the end of the run).
- ❑ "ρ" is the density of the liquid.
- ❑ "A" is the angle of the slope (if using a graphing calculator, be sure your settings are in degrees, not radians).
- ❑ "V" is the average velocity of the flow, which can be determined by dividing 10 cm by your run time (in seconds).

Use the equation exactly as written. Because the equation is designed for the units used in this activity, no unit conversions are necessary.

Worked Example

Suppose:

- ❑ h = 0.8 cm
- ❑ A = 15°
- ❑ V = 3.2 cm/s
- ❑ ρ = 0.92 g/cm³

$$\eta = \frac{980\ \text{cm/s}^2 \times (0.8\ \text{cm})^2 \times 0.92\ \text{g/cm}^3 \times \sin(15°)}{3 \times 3.2\ \text{cm/s}}$$

$$\eta = \sim 15.57\ \text{poise}$$

Tips for Success:

- ❑ Change only one variable at a time.
- ❑ Keep the pour rate as constant as possible.
- ❑ Measure temperatures immediately before each trial.
- ❑ Record measurements immediately after each run.
- ❑ Clean the ramp thoroughly before changing liquids.
- ❑ Use the same ramp, trough, and flow distance for all trials to ensure consistent results!

Initial Setup

Instructions

1. Create your trough
    a. Fold a strip of aluminum foil into a trough approximately 12-15 cm long, 3-4 cm wide, and 2-3 cm deep.
    b. Mark a start point and an end point on your trough, 10 cm apart.
    c. Fold the lower end upwards about 2 cm to create a stop that collects the liquid.
2. Build your ramp
    a. Construct your ramp on a smooth, rigid surface, such as a clipboard, baking sheet, or piece of foam board, approximately 25–40 cm (10–16 in) long.
    b. Raise one end using books or a box.
3. Place a foil trough on the ramp.

6

Part 2. Experimental Trials

1. Select a liquid and record its name in Table 6.1.
2. Using a protractor or inclinometer, measure the angle of your ramp and record it in Table 6.1.
3. Measure the temperature of your liquid and record it in Table 6.1.

Table 6.1. Measuring Viscosity

Trial #	Liquid (substance name)	Density[1] (g/cm3)	Temp-erature (°C)	Travel Time (seconds)	Velocity (cm/s)	Fluid depth (cm)	Angle of slope (degrees)	Viscosity (poise)	Notes
1									
2									
3									
4									
5									
6									
7									
8									
9									
10									
11									
12									

1. Get values from the Appendices (or your instructor).

4. Using a graduated cylinder or beaker, continuously pour approximately 20–30 mL of liquid onto the start line while another group member starts the stopwatch. Maintain a steady pouring rate until all of the liquid has been dispensed. Avoid sudden surges or pauses.
5. Measure the maximum thickness (depth) of the liquid at the front edge ("toe") of the flowing liquid immediately after it reaches the 10 cm mark. The timing can be tricky, so a third student should be ready to go with the ruler. Record this value in Table 6.1.
6. Repeat these steps for each of the other liquids, adjusting the temperature and ramp angle. Change only one variable at a time (substance, temperature, or ramp angle) while keeping all other conditions constant.

Hints:

- ❑ Keep trials of the same substance in consecutive rows. This will make it easier to compare the influence of the adjusted variables (temperature and ramp angle).
- ❑ Use ramp angles between 10° and 30°. Angles outside this range may cause liquids to flow too slowly or too quickly for accurate measurements.
- ❑ Complete approximately twelve trials in total. These may include different fluids, different temperatures, and/or different ramp angles, depending on your investigation.

6

7. Use the data collected in your table to calculate the viscosity of each liquid (last column).
8. Answer Activity 1 questions before moving on to Activity 2.

1. Which liquid had the lowest and highest viscosities? Support your answer using your calculated values.

2. Based on your data, explain the effect of *temperature on viscosity.*

3. Did changing the ramp angle change the actual viscosity of the fluid, or did it only change how quickly the fluid flowed? Explain.

Part 3. Gases in Liquids

Here, we will examine how gases behave in liquids of varying viscosities. Specifically, we will look at bubble shape and escape velocity as they rise through the liquids.

Tips for Success

- ❑ Use only transparent (or translucent) liquids.
- ❑ Practice timing before collecting recorded data.
- ❑ If using a smartphone video, record in slow motion when possible.

Trial Runs

Instructions

For each trial run:

1. Select the three fluids you will use for this experiment and observe their physical characteristics.

> Note: It will be important that the fluids you select are transparent or translucent.

2. Fill three graduated cylinders with each of the three fluids to the same marked height.
3. Use a syringe or dropper to inject a single air bubble (5-10 mm in diameter) at the bottom of the container (or as deep as your dropper can reach).
4. Make a few non-recorded/non-timed practice bubble releases and observe bubble shapes (e.g., spherical, oblong, conical), any changes in size, and paths of escape (e.g., straight, meandering) as they move through the liquid. Record your observations (including a bubble sketch) in Table 6.2.
5. Determine bubble velocity:
    a. Measure travel distance: measure the distance from where the bubble is released (at the bottom of the container) to where it meets the surface. Note that the surface will rise from its original height once the dropper is inserted! Record in Table 6.2.
    b. Measure the travel time of the bubble rise (from the moment of release at the tip of the syringe until it reaches the surface). This can be recorded using a stopwatch. A Smartphone video recording (slow-motion video if possible) may be needed for faster-moving bubbles.
    c. Calculate velocity. Record in Table 6.2.

> Note: If the bubble contacts the wall of the graduated cylinder during its ascent, discard that trial and repeat it.

Table 6.2. Gases in Liquids

Liquid #1: ____________________

Fluid 1 Trial #	Distance traveled (cm)	Time of travel (sec)	Velocity (cm/s)
1			
2			
3			
4			
5			
6			
Average velocity:			

Bubble shape and sketch
Bubble motion observations

Liquid #2: ____________________

Fluid 2 Trial #	Distance traveled (cm)	Time of travel (sec)	Velocity (cm/s)
1			
2			
3			
4			
5			
6			
Average velocity:			

Bubble shape and sketch
Bubble motion observations

Liquid #3: ____________________

Fluid 3 Trial #	Distance traveled (cm)	Time of travel (sec)	Velocity (cm/s)
1			
2			
3			
4			
5			
6			
Average velocity:			

Bubble shape and sketch
Bubble motion observations

6. Repeat 4–6 trials.
7. Repeat steps 2- 6 for the other two liquids.

4. Based on your observations of bubble behavior (not your calculated viscosities), rank the three liquids from lowest to highest viscosity.

5. What was your average bubble escape velocity for each liquid tested? List below.

6. Based on your observations, describe how viscosity influences the velocity of bubble escape.

7. Explain why bubbles changed shape more readily in highly viscous liquids than in low-viscosity liquids.

8. Based on your observations, describe how viscosity influences the paths the bubbles took during their escape.

6

Part 4. Bringing it all Together: Viscosity of Gases in Magma!

Scientific models are simplified representations of natural systems. Although the fluids used in this activity are much simpler than magma, they allow us to investigate the physical relationships among viscosity, gas movement, and volcanic eruption style.

9. How might the viscosity of a lava flow change with distance from the vent (as the lava moves further from where it erupted)? Explain your answer using the observations you have made here.

6

10. Which fluid(s) that you tested best represent low-viscosity basaltic magma? Which represents high-viscosity rhyolitic magma? Support your choices using evidence from both Activity 1 and Activity 2.

11. As magma rises toward Earth's surface, pressure decreases. Using Boyle's Law and the results from Activity 2, explain what happens to gas bubbles during magma ascent.

12. How might rapid gas escape from low-viscosity magma influence both the style of eruption and the types of volcanic hazards produced?

13. If gas bubbles cannot escape efficiently from magma, describe what happens to:

- ❑ pressure inside the magma
- ❑ bubble size
- ❑ eruption style

14. Mount St. Helens has erupted both explosively (1980) and effusively (2004–2008). Based on what you learned in this lab, explain why these eruptions were so significantly different.

15. Using evidence from this lab activity, explain why volcanoes with similar magma composition can still produce very different eruption styles.

6

6

16. Discuss some potential sources of error in these experiments (excluding simple recording or calculation mistakes). Which factors might still affect the reliability and accuracy of our results?

17. Based on your observations, describe how viscosity influences the paths the bubbles took during their escape.

Name: ______________________________ Lab Time: __________________ Due: ____________________

7 Introduction to Google Earth Pro

Purpose

In this lab, you will learn how to use Google Earth Pro and how to extract basic information about topography, landscapes, rock formations, and geologic hazards.

Objectives

You will learn how to navigate in Google Earth to view a great variety of geologic surface features. The several labs will require these Google Earth skills:

- ❑ Input latitude and longitude and get to a location of interest.
- ❑ Zoom in and zoom out.
- ❑ Learn to turn layers on and off.
- ❑ Control oblique view.
- ❑ Save an image/screenshot and drop it into a Word document.
- ❑ Extract the position and elevation of a point.
- ❑ Use the ruler tool to measure distance along a straight line or a curving path.
- ❑ Use the trace tool to identify or highlight a feature.
- ❑ Make and save a path.
- ❑ Make and save a placemark (pin).
- ❑ Create a topographical profile on land.
- ❑ Create a topographical profile on the sea floor.
- ❑ Use historical satellite imagery.

Background

Google Earth Pro[1] is a free software application that connects to a large Google database and interactively serves images of portions of our Earth. The images are a stitched-together, complex combination of aerial photos, satellite images, and surface topography models at different resolutions. There are different layers with different types of information that you can turn on or off to customize what you want displayed in the image.

---

1. A web-based Google Earth version also exists but lacks many of the functions needed to complete the Google Earth lab exercises for this course. The mobile version lacks even more. Please don't use these to complete your labs.

Google Earth has features similar to traditional maps. This includes an underlying projection with positions referenced by latitude and longitude or UTM, a scale bar, and representations of geographic features such as roads and political boundaries. It also allows viewing of 3-dimensional features. If you are interested in a fuller, more technical description of Google Earth, it can be found on Wikipedia using QR Code 7.1.

QR Code 7.1. Technical Description of Google Earth on Wikipedia

Read This: Essential Functions of Google Earth Pro

Before you get started, open the user guide using QR Code 7.2 and read over the sections in the Essential Google Earth Pro Functions.

QR Code 7.2. User Guide for Google Earth Pro

The essential functions you should focus on are:

- ❑ The Google Earth Pro User Interface
- ❑ The Google Earth Pro Toolbar
- ❑ Navigation
- ❑ The Google Earth Pro Status Bar
- ❑ Searching
- ❑ Working with Places
- ❑ Creating a New Placemark
- ❑ Drawing Paths and Polygons

> Note: Don't stress if you didn't understand everything that you just read. Just having a basic understanding of these functions will make your work easier.

Part 1. Start Here

Instructions

1. First, download Google Earth Pro if you do not already have it.
2. Google search "Download Google Earth Pro," then follow the prompts. This works best on a desktop or laptop (as opposed to a mobile phone). Google Earth Pro cannot be downloaded onto a Chromebook, iPad, or device with limited processing capacity.
3. Go to the 'View' option at the top of the screen as shown in Figure 7.1. Make sure all these options are checked.

Figure 7.1. Click on the "View" menu in Google Earth Pro.

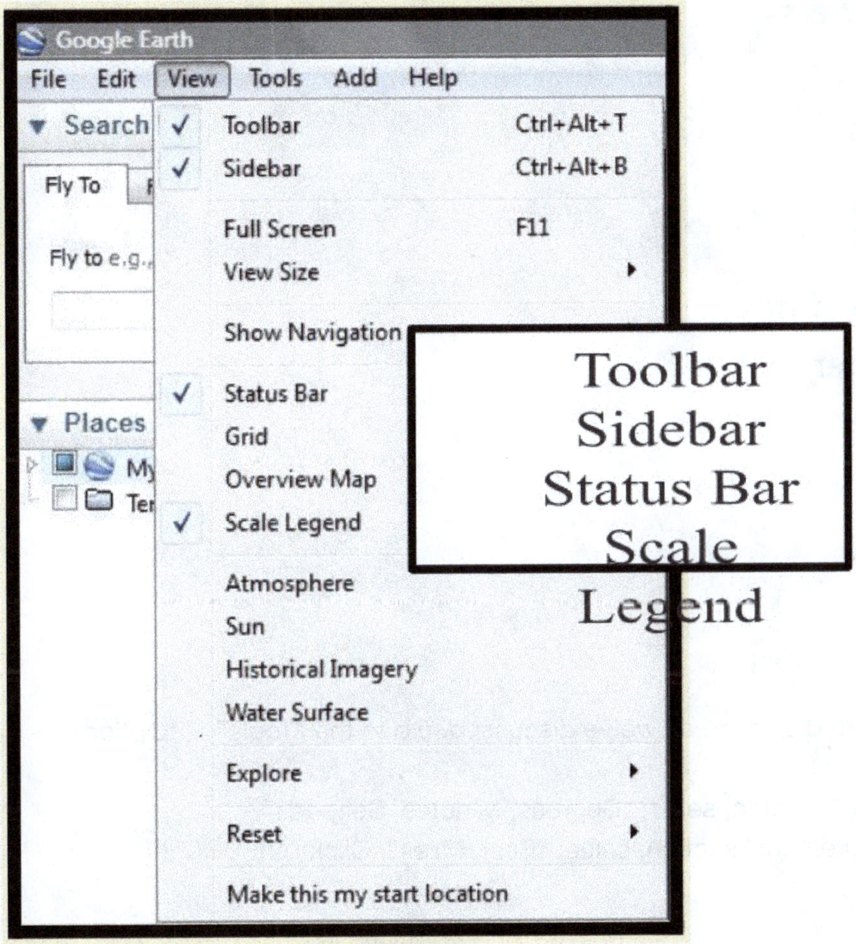

4. Next, we want to minimize map clutter. Uncheck most of the options under "Layers," especially "Places" and "Photos." Keep "Borders" but remove "Labels."
5. Make sure "Terrain" is selected as shown in Figure 7.2.

Figure 7.2. In the "Layers" menu, select only "Places," "Borders," and "Terrain."

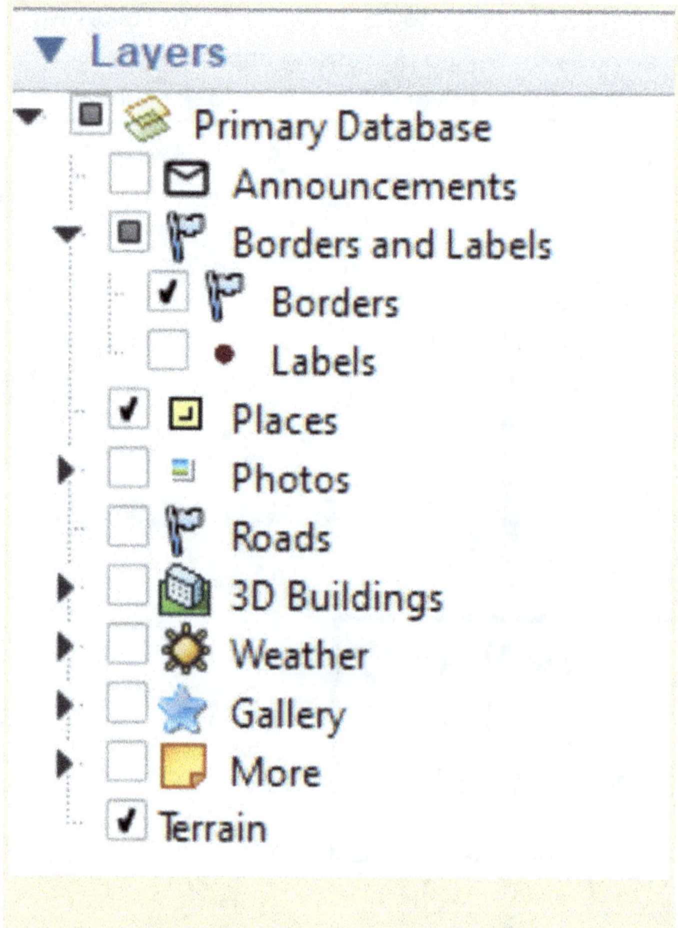

6. Make sure the units are displayed as we've discussed. Go to the "Tools" > "Options" > "3D View" tab.
7. In the "Show Lat/Long" section, select "Degrees, Minutes, Seconds"
8. In the "Units of Measurement" section, select "Feet, Miles". Click "OK".

Spend time familiarizing yourself with how to move and adjust the map view using the navigational controls in the upper-right corner (shown in Figure 7.3), including zooming in and out and moving the map around.

Figure 7.3. Google Earth Pro navigation tools.

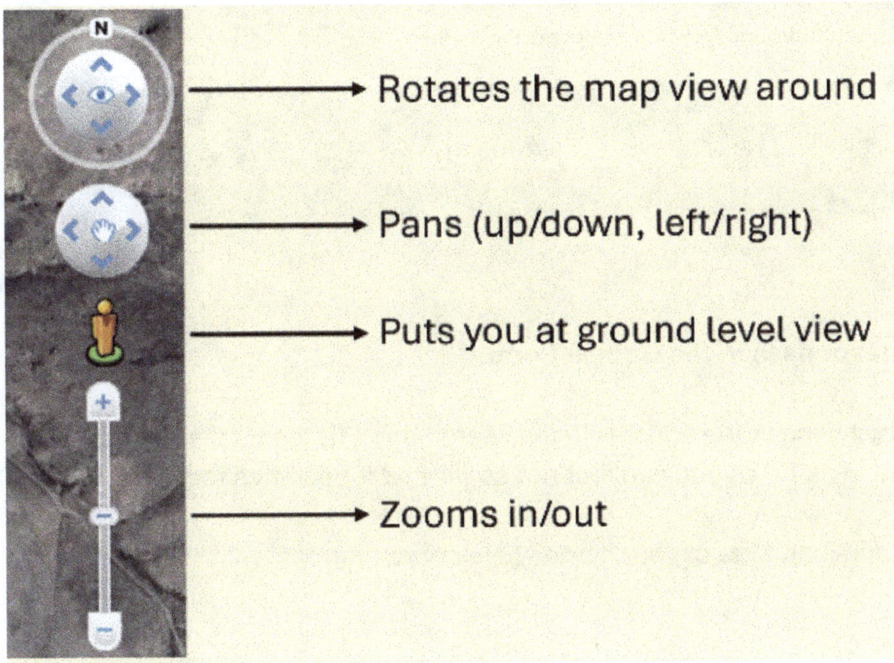

9. Zoom in and out with your mouse wheel, touchpad, or zoom slider bar.

> Pro tip: the letter "R" is your friend! If things get weird, just click on the map and type the letter "R". This will make everything right—meaning North is up, and you're looking at the land in map [overhead] view.

Part 2. Determining Latitude, Longitude, and Elevation

Places where you need to answer a question or insert an image are highlighted in grey. Answer the questions in the corresponding Google Form. Unless otherwise indicated, use feet for elevation and water depth and miles for distance.

Latitude, Longitude, and Elevation of Your Home

Latitude, longitude, elevation, and a few other attributes of a place on Earth are shown at the bottom right of your screen, and the values change as you move your cursor over the map area, as shown in Figure 7.4.

Figure 7.4. Example of Latitude, Longitude, and Elevation in Google Earth Pro.

Instructions

Enter or upload all your responses in the Google Form.

1. "Fly to" your home by typing in the address in the search bar. Put the cursor directly over your home and zoom in as far as you can (but not so far that it becomes too pixelated).

1. What is the latitude and longitude of your home? (use correct format).

2. What is the elevation of your home (in feet)?

3. What is the imagery date of the location of your home?

Measuring Distances

Measuring distances in Google Earth can be done with several different tools, but the simplest is the ruler.

Instructions

Enter or upload all your responses in the Google Form.

1. While you're looking at your home, measure the width of the street you live on.
2. Select the Ruler tool, and a Ruler box will appear. Make sure that feet are selected.
3. Click first on one side of the street, then directly across to the other side. A yellow line will appear, and the length of the line you made will show up in the Ruler box.

4. What is the width (length measurement) of your street in feet?

5. What is the width (length measurement) of your street in another unit (your choice!)

6. What is the straight-line distance from your home to Chemeketa Community College (Salem Campus) in miles?

Topography of the U.S.

Topography is the shape of the land surface, which is determined by the distribution of elevation across an area.

Instructions

Enter or upload all your responses in the Google Form.

1. Zoom out to where most of the contiguous U.S. is shown (lower 48 states only, sorry Hawaii and Alaska!)
2. Move your cursor over the display/map part and observe how the elevation changes as you move over various parts of the contiguous United States.
3. Review how it is related to features such as mountains, plains, river valleys, coasts, etc.
4. Use Figure 7.5 and think about how topography/elevation is correlated with the different U.S. geologic provinces. Spend some time doing this.
5. Move your cursor around and search for the lowest elevation on land that is *not* adjacent to an ocean.

7. What is it (in feet) and where is it? (indicate general part of specific state and give the elevation).

8. What U.S. Geologic Province is it in?

6. Move your mouse around and search for the highest elevation on land.

9. What is it (in feet) and where is it? (indicate general part of specific state and give the elevation). What U.S. Geologic Province (from Figure 7.5) is it in?

Figure 7.5. U.S. Geological Provinces.

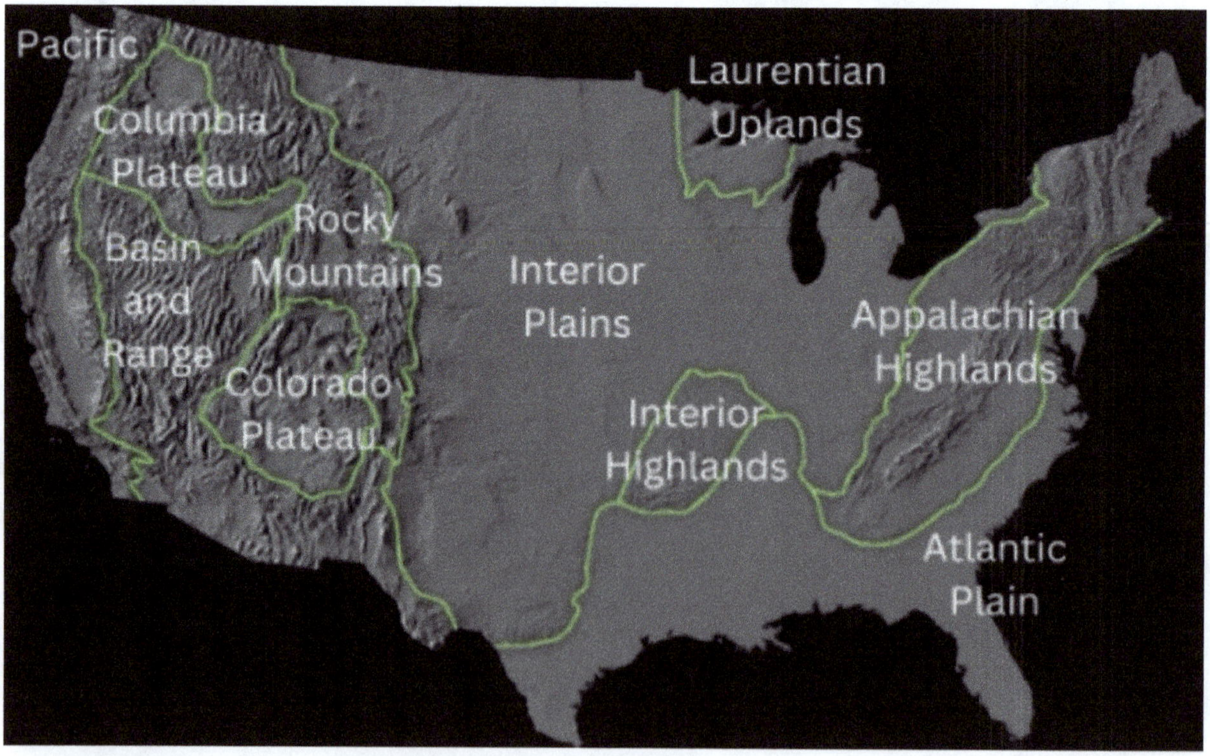

Water Depth

As with elevation, water depth can be determined by moving your cursor over a body of water and observing the value displayed in the lower-right corner. The depth values will be negative since the seafloor is below sea level.

Instructions

Enter or upload all your responses in the Google Form.

7. Move your cursor offshore (Pacific or Atlantic is fine) and observe how the values change with distance from land and over different seafloor features.

10. Describe how the color of the seafloor is related to depth.

8. Find the location of the Titanic Shipwreck (use the search bar). Zoom in as close as you can.

11. What is the water depth of its cold, dark resting place? What is the imagery date?

Elevation Profile Across the U.S.

Earth scientists LOVE elevation profiles, which are 2-dimensional cross-sectional views of a landscape that highlight the land's shape along a specific line. Examples are shown in Figures 7.7 and 7.8. This is a necessary skill for this course.

Here you will create a path across the U.S., from the northern California coast to the Maryland/Virginia coast, and get a sense of the topography along a long stretch of the country.

Instructions

Enter or upload all your responses in the Google Form.

1. Select the "Borders and Labels" box in the Layers menu. (This will help you see the state lines and a few other things to help you navigate.)
2. Select the Add Path tool.

Start Your Path

3. Click first on the west coast starting point (northern California coast – right at a beach on the Pacific Ocean), then the east coast ending point (Maryland/Virginia area – at an Atlantic Ocean beach).

> Hint: Keep your profile line more or less straight. Try to minimize "jumping around" as much as possible. If needed, you can adjust the points when they are highlighted in blue.

4. Give your path a name (e.g., "West to East Coast Profile") and click "OK".
5. Right-click on it in the "Places" menu. Click "Show Elevation Profile".
6. Run your mouse along the elevation profile at the bottom and notice how the elevation changes along your profile line. A marker arrow will show you the ground location that corresponds with your cursor's position on the elevation profile.
7. Notice that at the top of the profile, some very useful numbers are displayed – elevation minimum, maximum, and average, total length of your profile (distance), slope values, etc.
8. Take a screenshot of the profile image and the profile line on the map insert it in the Google Form.

7

> Hint: paste your copied image into a Word document. You can upload the Word doc as is or save the image as an image file by right-clicking on it, then saving it as a picture (once it is pasted into your Word doc). Make sure the full profile line on the map is visible, along with your elevation profile.

Figure 7.6. Example of an elevation profile in Google Earth. The red line on the map is the profile line, and the graph at the bottom is the elevation profile (also known as a topographic profile). Snake River Canyon at the Oregon-Idaho border.

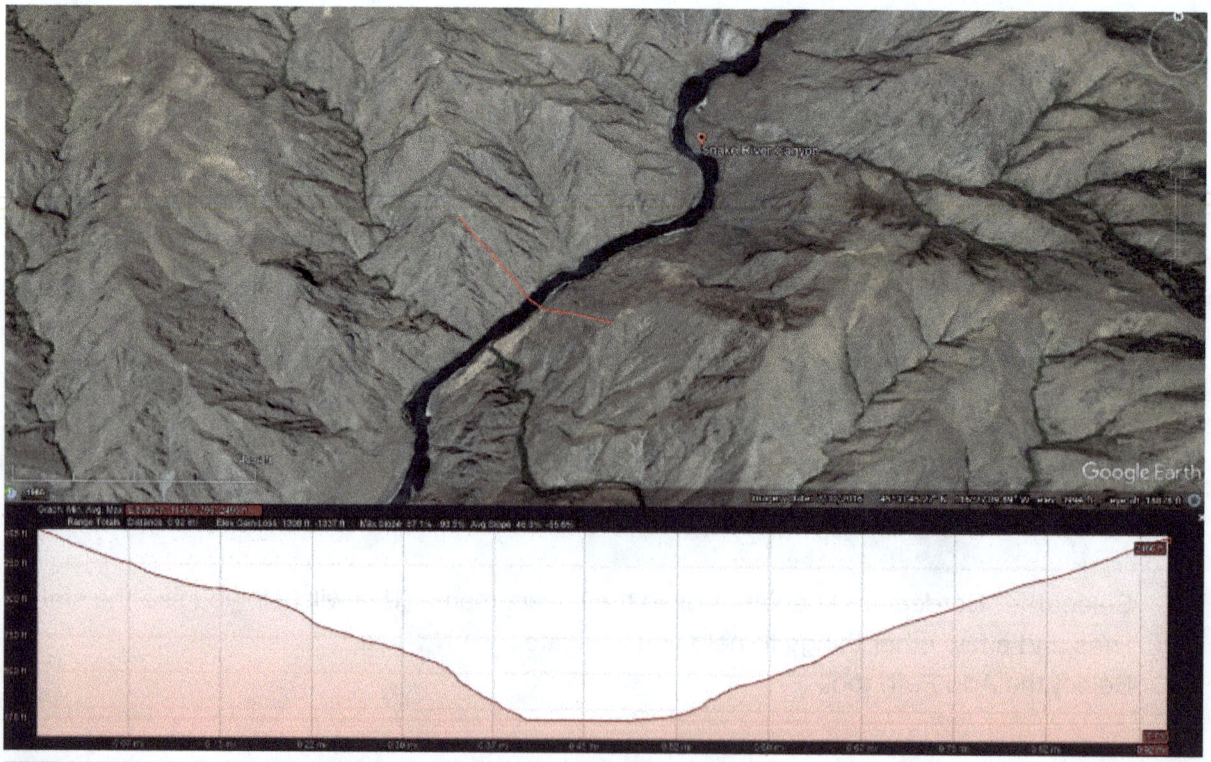

Figure 7.7. Example of an elevation profile in Google Earth. The white line on the map is the profile line, and the graph at the bottom is the elevation profile (also known as a topographic profile). Chief Kiawanda Rock (also known as Haystack Rock) off Pacific City, Oregon.

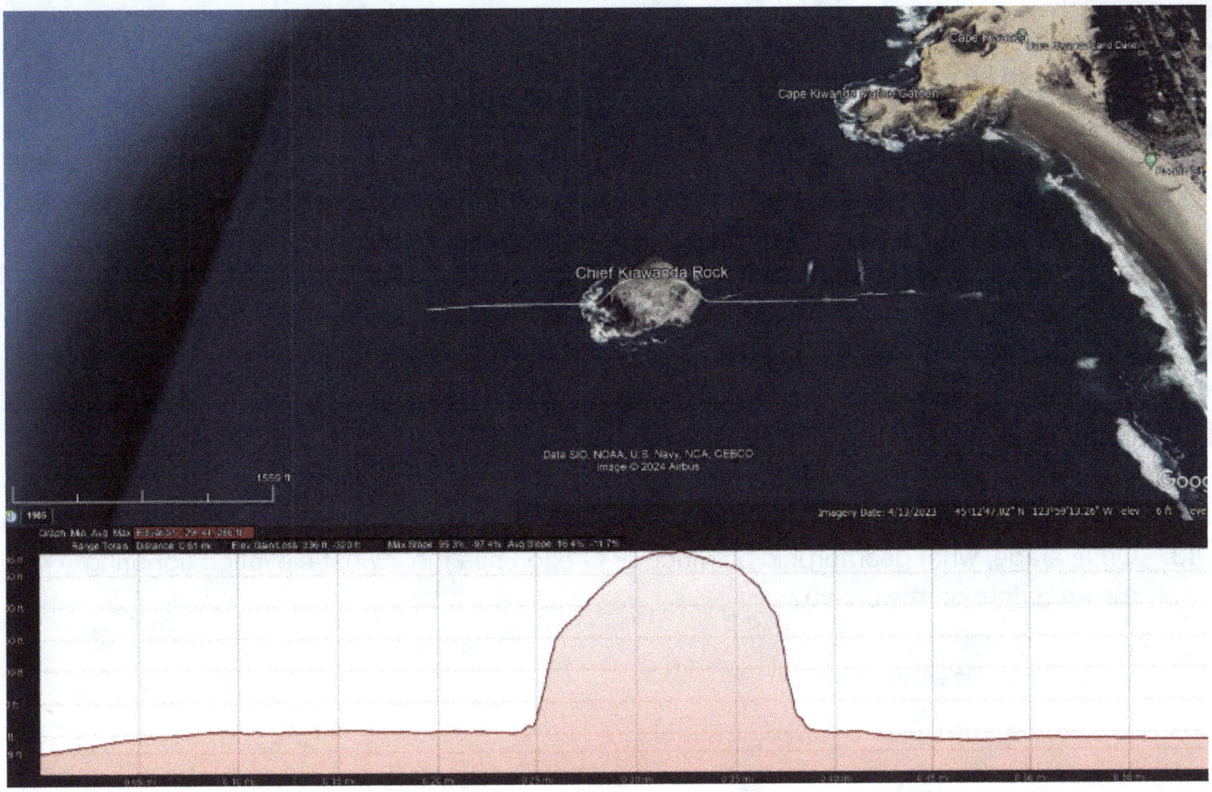

12. Describe your elevation profile.

How long is it?

How do elevation and topography change as you move from west to east? Be sure to include the relevant provinces, states, and mountain range names in describing this change. This should be at least 4 sentences.

Elevation Profile Across Oregon

Same elevation profile, but across Oregon!

Instructions

Enter or upload all your responses in the Google Form.

1. Navigate to Oregon and select "Add Path" (as you did in the previous question) and create a line across northern Oregon from WEST to EAST (more or less straight) that starts right at the coast, runs through Mt. Hood, and ends at the Idaho border.
2. Give it a name before you click OK.
3. In "Places", right-click on the name of the cross-section you just made and select "Show Elevation Profile". This will appear below.
4. Run your mouse along the elevation profile at the bottom and make some observations about how the elevation and relief change.

7

13. On average, what geomorphic province of Oregon has the lowest elevation (ignoring the area right on the coast)?

- ❑ Highest elevation?
- ❑ Steepest terrain?
- ❑ Flattest terrain?

Refer to Figure 7.8.

Figure 7.8. Oregon Geological Provinces.

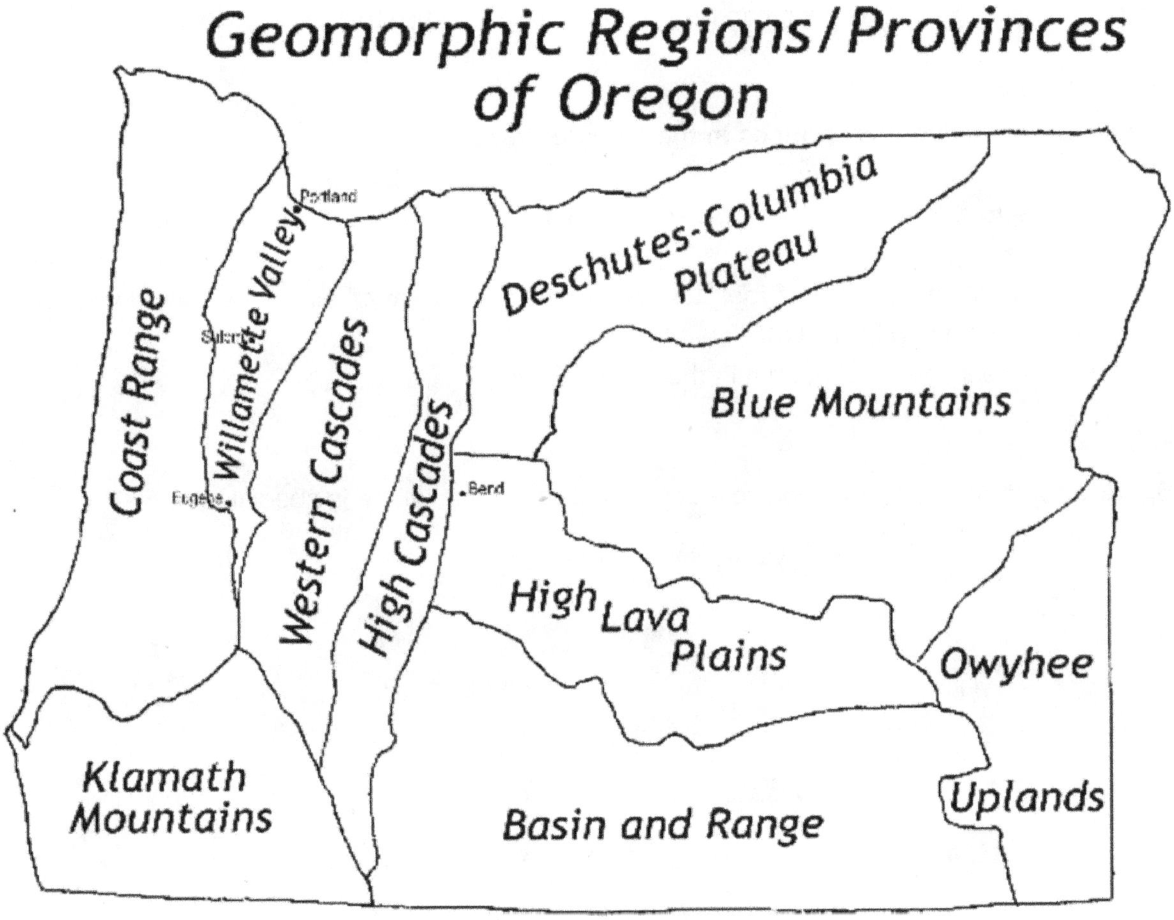

After D.M. Johnson et al., Atlas of Oregon Lakes, *1985, and A.M. Tattersall, 2005*

14. What is the total distance of your profile (in miles)? Information is shown above the profile graph.

5. Take a screenshot of the profile image with the map and profile line visible and insert it into the Google Form.

Placemark a Location

Placemarks are digital "pins" that can be saved to quickly return to a specific location.

Instructions

Enter or upload all your responses in the Google Form.

1. Navigate to a place you have a personal attachment to (e.g., Grandma's house, Branson, Key West)
2. Select the "Add Placemark" tool and move the displayed pin to the exact location.
3. Give it a name and click "OK".
4. Take a screenshot and insert it in the Google Form.

15. Say something about this place – describe it or explain why it is special to you (optional).

Historical Imagery

Satellite images of places are constantly updated to reflect changes in the land surface due to natural causes (erosion, volcanic eruptions, landslides, fires, etc.) and anthropogenic causes (land-use changes, construction, etc.).

Google Earth displays the most recent images, but stores older ones too, which can be viewed as far back (generally) as the mid-1980s.

Instructions

Enter or upload all your responses in the Google Form.

1. While you're still at your "special place" from the previous question, select the "Historical Imagery" tool from the "Toolbar." Notice how a slider bar appears.
2. Move the slider to different dates and observe how the land surface has changed over the last few decades.

16. What is the year of the oldest image provided for your location?

17. Describe two specific changes you notice as you move forward in time to the present.

Note: the earliest image might be too blurry to comment on. Use the first useful image as a starting point.

Polygons!

A polygon is a closed, two-dimensional shape that is made up of at least 3 line segments (e.g., triangles, squares, pentagons, etc.) In Google Earth, a polygon can be used to measure a specific area or simply highlight it.

Instructions

Enter or upload all your responses in the Google Form.

1. Navigate to a city park near you (or one you currently enjoy or have enjoyed in the past). Zoom in as far as possible (while still being able to view the whole thing) and center it. Select the "Add Polygon" tool from the Toolbar and click around the margins of the park as carefully as possible (this may take some practice).
2. In the "New Polygon" box that appears, click the "Style, Color" tab and change the area's color to your choice (not white).
3. Change the opacity to 50%.

NOTE: If you already closed out of the "New Polygon" box, you can get back to it by right-clicking on its name (in the "Places" menu) and then selecting "Properties".

4. In the "Measurements" tab, determine the area of the park in square feet and acres. Enter your response in the Google Form.
5. Give it a name and click OK.
6. Instead of taking a screenshot, we will save an image this time.
7. Click the "Save Image" tool in the Toolbar. Give it a name by clicking on the box that appeared in the upper left corner of the map area.
8. Click the "Save Image" button that is now beneath the Toolbar. Insert this image into the Google Form.

7

Name: ______________________ Lab Time: ______________ Due: ______________

8 Igneous Rocks and Volcanoes of the Pacific Northwest

Purpose

In this lab, you will use Google Earth Pro to study igneous rocks and volcanoes of Oregon and other parts of the world.

Part 1. Start Here

Follow the directions below before answering the questions.

> Note: The first 4 steps below may be redundant with Lab 7, since your settings have probably not changed, but it is good to double-check. Don't overlook the last 2 steps!

Instructions

1. Open Google Earth Pro (or download it if using a new device; see Lab 7).
2. Go to the 'View' option at the top of the screen. Make sure all these options are checked.

Figure 8.1. Click on the "View" menu in Google Earth Pro.

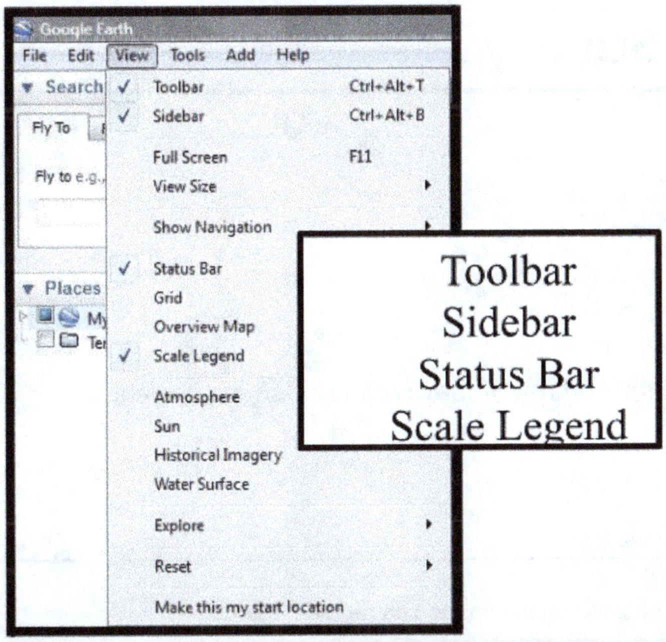

3. Declutter the map: uncheck most of the options under “Layers”, especially “Places” and “Photos”. Keep “Borders”, but remove “Labels”. Make sure “Terrain” is selected.

Figure 8.2. In the “Layers” menu, select only “Places,” “Borders,” and “Terrain.”

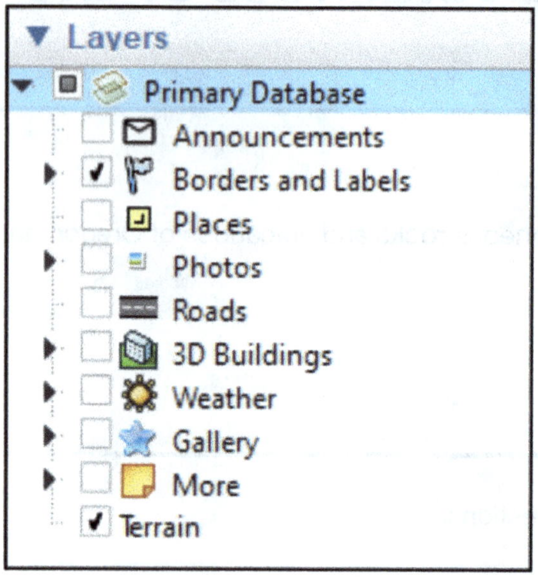

4. Make sure that latitude and longitude are displayed in degrees, minutes, and seconds, that units of measurement are feet and miles, and that the measurement units are English. Go to “Tools > Options.”
5. Download the “kmz” file (found in Canvas module) to your device.
6. Open the “kmz” file in Google Earth by clicking “File>Open” and selecting the “kmz” file. This will put placemarks (pins) on your map.

Part 2. Volcano Type Survey

Mt. Hood, Oregon

Instructions

Enter all your responses in the Google Form.

1. Navigate to this location using the search bar.

1. What is the latitude and longitude[1] of the summit of the volcano? Use the correct format.

1. Remember: the latitude/longitude of the cursor position on the image can be read from

2. Find an oblique[2] view - using the navigational controls (or just zoom in and adjust, or shift+scroll mouse), find an oblique view that shows the full form of this volcano (the whole thing!) See the example in Figure 8.3.

Figure 8.3. Example of an oblique view (Mt Rainier, Washington).

3. Take some basic measurements of Mt Hood.
4. Draw a path that bisects the volcano. It should start at the base of the west side, travel over the peak (right through the summit), and back down to the edge of the base on the east side. It should extend to the relatively flat lands on both sides of it (for a total length of about 15-20 miles) to make sure you get the whole thing – it is much bigger than the "tree line" or snow line, which are usually the most obvious features you can see from above.

> Hint: you can adjust a saved profile line by right-clicking on it in the "Places" menu, then selecting "Properties". Click the point you want to adjust (it will highlight in blue, then green), and move it accordingly. It should look roughly like the example shown in Figure 8.4.

---

2. An oblique view is between aerial (overhead/map) view and on the ground ("street view").

Figure 8.4. Example of a profile line that approximately bisects Mt Rainier. Elevation profile shown below. Note that the slope percentages can be positive or negative (negative just means that it is calculated going downhill along the profile line). A maximum slope can be negative.

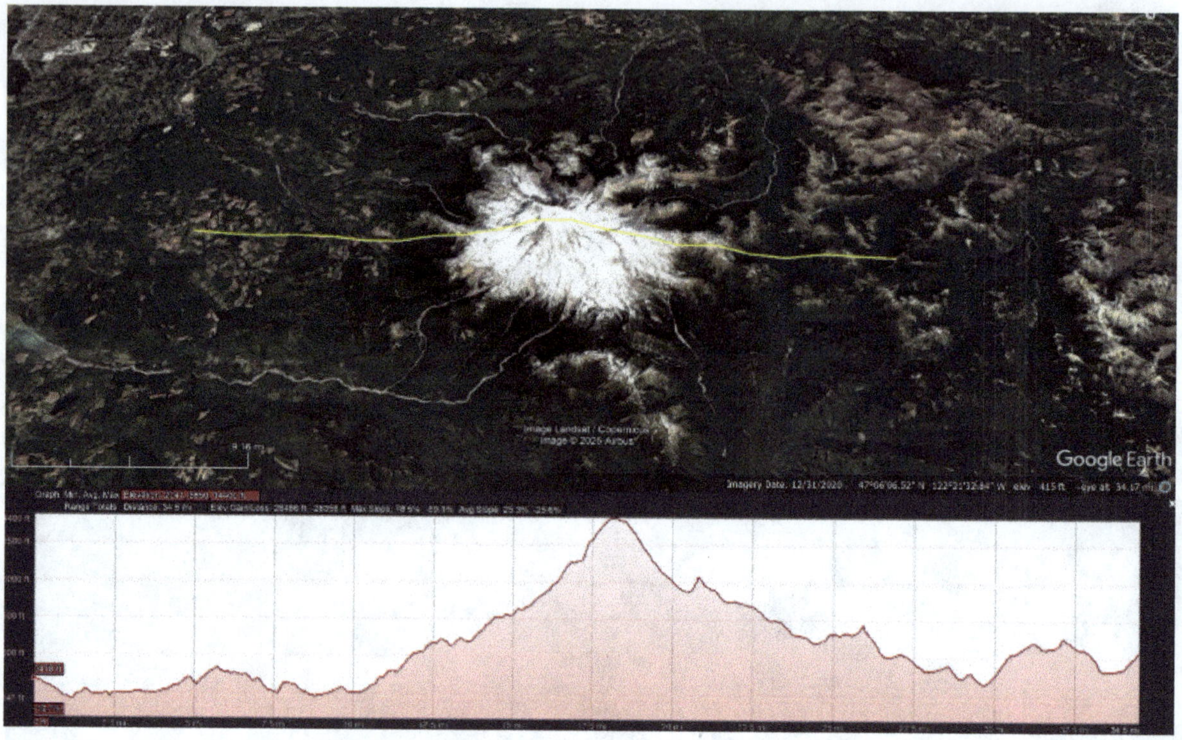

5. Name your path "Mt Hood."
6. Make an elevation profile using the same process as in Lab 4, and observe how the slope and elevation change as you move your cursor along the profile.
7. Take a screenshot showing the volcano's full form in oblique view, the profile line, and the elevation profile. Insert it into the Google Form.

> Reminder: Enter all your responses in the Google Form.

2. What is the maximum % slope (positive or negative) shown in your profile?

3. Describe where approximately on the mountain the highest slope values and the lowest slope values are found.

> Hint: on the elevation profile, you can highlight a specific section of the profile (click and drag), and slope values will automatically be recalculated for only the highlighted section. Negative slope values just mean that they are calculated going "down" the profile. If the highest numerical value of the slope is negative, use that one.

4. What is the height of Mt Hood as measured in your profile (in feet)?

- ❑ To determine this, place the cursor over the highest point on Mt. Hood and note the elevation.
- ❑ Place the cursor on a point some distance away from the peak where the slope of the volcano is no longer evident (approximate) and note the elevation there.
- ❑ Subtract the two values.

Note: It is not asking about the summit elevation, but rather how high the mountain stands above the surrounding landscape.

5. How wide is the base of Mt Hood? Estimate this by highlighting a section of the profile as described above. Remember: the base of the volcano (where it "starts") is where you start to see a significant increase in slope.

6. What type of volcano is this? What evidence shown in Google Earth supports this? Explain.

8

8. Return to aerial (overhead) view. Remember, this can be done simply by clicking on the map and typing the letter "R".

Capulin Volcano National Monument, New Mexico

Instructions

Enter all your responses in the Google Form.

1. Navigate to this location using the search bar.

7. Determine the latitude and longitude of the center/summit of the volcano. Give your answer in the correct format.

2. Create an elevation profile across the volcano, about 2 miles in total length. Use the same steps as above (west to east across the feature).
3. Take a screenshot that shows an oblique view of the entire volcano, a profile line on the map, and an elevation profile, and insert it into the Google Form.

8. What is the maximum % slope (positive or negative) shown in your profile?

9. What is the height of Capulin as measured in your profile (in feet)? Calculate using the same process as described above.

10. How wide is the base of Capulin (in feet or miles)? Measure using the same process as described above.

11. Capulin last erupted around 60,000 years ago as a lava flow. Based on the appearance of the landscape that surrounds it, in what direction did the lava mostly flow (and how can you tell)?

12. What type of volcano is this? What evidence shown in Google Earth supports this? Explain.

Hawaii (Big Island)

Instructions

8

Enter all your responses in the Google Form.

1. Copy and paste the following coordinates into the search bar: 19°49'19" N, 155°28'02" W.
2. Once you get there, zoom way out so that you can see the island (Hawaii/the Big Island) and the surrounding waters.
3. Make an elevation profile for this volcano that extends beyond both sides of the island and well into the ocean, going as close to straight west to east as possible.
4. Make your path begin and end in the ocean on opposite sides of the volcano, but going through the given coordinates. It should be about 160 miles across with the island right in the middle.

This volcano rose from the seafloor rather than atop a continent, making sea level essentially irrelevant. We want to look at it from the seafloor instead.

5. Right-click on the path again and select Properties. Then click the "Altitude" tab and change from "Clamped to ground" to "Clamped to seafloor". Notice how your profile changed.

13. What is the name of the volcano that you are profiling?

14. What are all the relatively small, round landforms that exist near the summit of the volcano/mountain?

6. Take a screenshot showing an oblique view of the entire volcanic island, a profile line on the map, and an elevation profile.

15. What is the maximum % slope (positive or negative) shown in your profile?

16. How tall is the volcano from base (seafloor) to peak (in feet)?

17. How wide is the base of this volcanic island (in miles)? Measure using the same process in previous questions, but note that this will extend into the Pacific Ocean on both sides.

18. What type of volcano is this? What evidence shown in Google Earth supports this? Explain.

8

Volcano Type Wrap-Up

Instructions

Enter all your responses in the Google Form.

19. Now that you have seen a few volcanoes, describe in 1-2 full paragraphs the physical characteristics of these volcanoes and how they relate to volcanic processes and plate tectonics. Based on their overall sizes, slopes, crater characteristics, and locations, what can you deduce about their formation and history?

Note: Some questions to think about when you write your paragraphs could include but aren't limited to:

- ❑ How did the overall sizes vary?
- ❑ How do the slopes vary?
- ❑ How do the crater characteristics vary?
- ❑ Which volcanoes have a history dominated by explosive vs effusive eruptions?
- ❑ What do you know about the various tectonic locations where they are found?

Make sure that you compare and contrast these specific volcanoes.

Part 3. Recent Volcanic Events

Hawaii (Big Island)

Instructions

Enter all your responses in the Google Form.

1. Copy and paste the following coordinates into the search bar: 19°29'47" N, 154°49'13" W.
2. Keep this location centered and zoom out to see more of the eastern part of the island (to an "eye alt" of about 12,000 ft).
3. Observe the date of the satellite image shown at this specific location.

20. In historical imagery, go back to around October or December of 2011 (make sure you can see the island clearly, with minimal cloud cover). Carefully observe and describe the differences between the current and historical satellite photos at this specific location. What happened to cause these differences? Be specific.

4. In the historical image, find Kapoho Bay and roughly trace it by using the Path tool (multiple click points will be needed). Name the path "Kapoho Bay".
5. Change the appearance of this line by selecting the "Style, Color" tab. Change the color to bright red and the width to 2.0.

8

Hint: if you already closed out of the box, get back to it by finding the name in the Places menu, right-clicking on it, and selecting "Properties".)

6. Close out of the historical image and observe how the shoreline in this area changed.
7. Take a screenshot showing the Kapoho Bay path on the modern landscape.
8. Using the Ruler tool, measure the distance from the innermost part of (the former) Kapoho Bay to where the shoreline is currently.

21. How far (in feet) did the shoreline extend due to this event?

Part 4. Igneous Rocks, Volcanoes, and Structures of the Pacific Northwest

For several of the following questions, you will use the pins placed on your map when you opened the kmz file. Check the "Places" menu to ensure they are there and the boxes are selected. Click the triangles next to the file name to ensure the pins are shown in list view. This will make it easier to navigate through this part of the lab by double-clicking on them.

Mt. St. Helens, Washington

Instructions

Enter all your responses in the Google Form.

1. Navigate to this location using the search bar. Center over the crater and zoom in.
What type of volcano is this?

22. Describe the shape of the crater. What does it tell you about the blast (from the 1980 eruption) that formed it?

2. Look at the crater from as close to ground level as possible by holding the shift button and scrolling up with your mouse wheel.
3. Zoom in and look closely at the large "hump" in the middle of the crater.

23. What is this volcanic feature?

4. Do a 360° rotation around the volcano summit using either the direction wheel navigational control (or by holding the control button on your keyboard and scrolling with your mouse wheel). Try to get to the top of the crater and "stand" on the rim.
5. Notice the other prominent (mostly snow-covered) mountains in the background. Those are other Cascade volcanic peaks!

24. Which ones can you identify? Two of them should be very obvious. At least two others should be noticeable. Observe the compass direction and reference the Cascade peaks map in Figure 8.5 to help you determine this. Name at least 3 specifically.

Figure 8.5. Map of the most prominent Cascade peaks.

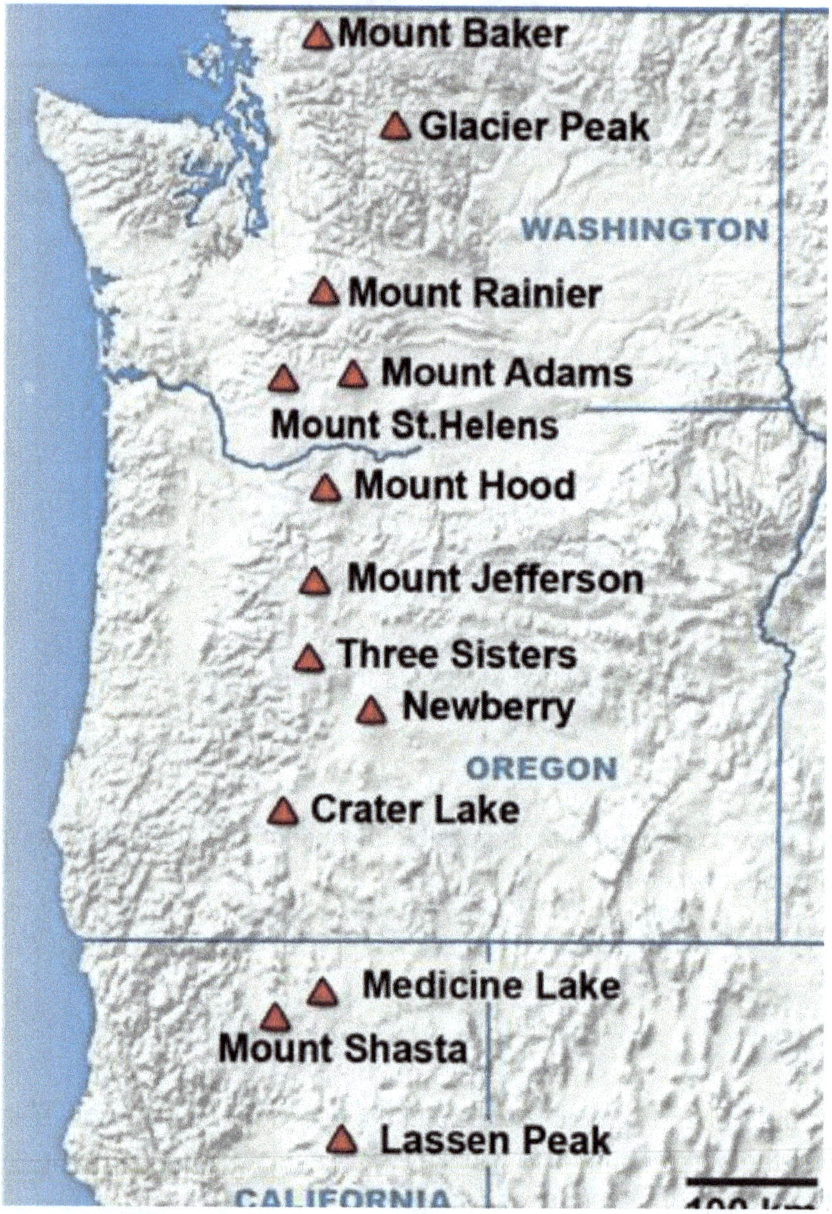

6. Make an elevation profile across the mountain, through the approximate middle of the crater, from west to east. Try to get your line to go right through the middle of the crater.
7. Look at your elevation profile and move your mouse until you find the two highest elevations representing each side of the crater on the crater rim.

25. What are these two elevations (in feet)?

26. What is the lowest elevation you see in the crater (along your profile)?

27. Based on your two previous answers, what is the relief of the crater? Reminder: this is the highest elevation minus the lowest.

8. Take a screenshot that shows an oblique view of the full volcanic form, the profile line on the map, and the elevation profile, and insert it into the Google Form.
9. Navigate to the large lake almost straight north of the crater.

28. What is the name of this lake?

10. Zoom way in to the north end of the lake, where the color is grayish-brown.

29. What do you see floating in the lake that is creating this color/feature? Why are they there?

Newberry Volcano, Oregon

Instructions

Enter all your responses in the Google Form.

1. Go to the Newberry Volcano pin.
2. Observe this feature and the surrounding landscape using both overhead and oblique views.
3. Make an elevation profile approximately west to east that goes through and approximately centers Paulina and East Lakes as well as the central cone between them. It should be 15-18 miles long.
4. Take a screenshot that shows an oblique view of the entire volcanic area, a profile line on the map, and an elevation profile, and insert it into the Google Form.

30. This location is generally considered part of the Cascade Range, but some geologists aren't so sure. Compare this location and landscape to the other Cascade volcanoes you've observed and describe at least 2 significant differences.

31. This volcanic area is associated with hundreds of cinder cones! Find at least three and encircle their perimeters using the "Path" tool. Choose your own color and make the weight 2.0.

32. Take a screenshot showing the cinder cones and insert it into the Google Form.

Mt Tabor, Oregon

Instructions

Enter all your responses in the Google Form.

1. Go to the Mt Tabor pin.
2. Observe this feature and compare it to the surrounding landscape.
3. Make an elevation profile west to east across this feature (but avoid the reservoir).

33. Based on your general observations and the elevation profile, what type of volcanic feature is this?

34. This is but one of many volcanic vents of the Boring Lava Field. Research this volcanic area and describe its geologic history in 3-4 sentences. Use your own words and be sure to include its age.

35. In which Oregon geological province is this located?

Wallowa Mountains, Oregon

Instructions

Enter all your responses in the Google Form.

1. Go to the Wallowa Mountains pin.
2. Zoom out to observe the overall topography, then zoom in to observe the peaks of these mountains, comparing them to the locations you looked at previously.

36. Based on their overall appearance, are these volcanoes? Explain.

37. In which Oregon geological province is this located?

Deschutes Canyon, Oregon

Instructions

Enter all your responses in the Google Form.

1. Go to the Deschutes Canyon pin.
2. Use your navigational control skills and look along the upper part (above the road) of the eastern canyon wall.
3. Observe the rock exposure and visible structures.

38. What type of rock is this?

39. What rock feature/structure is visible?

40. In which Oregon geological province is this located?

Leslie Gulch, Oregon

Instructions

Enter all your responses in the Google Form.

1. Go to the Leslie Gulch pin.
2. Look around at the exposed rock and landscape and compare it to what you observed at the previous location.

41. Describe the rock (especially color/composition) and compare it to the rock you observed at the previous location.

42. What type of rock is this?

43. In which Oregon geological province is this located?

Jordan Craters, Oregon

Instructions

Enter all your responses in the Google Form.

1. Go to the Jordan Craters pin.
2. Zoom way in and way out and look around at the exposed rock and landscape.

44. Describe the rock (especially color/composition) and compare it to the rock you observed at the two previous locations.

3. Get a close look at the rock exposure.
4. First, zoom way in (to an eye alt of about 5000 ft) and look closely as you move around.
5. Next, in the Layers menu, select "Photos" and view some of the pictures that people have shared in the area. What type of volcanic material is this?
6. Zoom out again and trace the area of this exposure using the Polygon tool.
7. Determine the area.

45. What is the area of Jordan Craters (in square miles)?

46. Research this location and describe its geologic history in 3-4 sentences. Use your own words and be sure to include its age.

47. In which Oregon geological province is this located?

Fort Rock, Oregon

Instructions

Enter all your responses in the Google Form.

1. Go to the Fort Rock pin.
2. Observe the feature in both overhead and oblique view.

48. Describe its appearance.

8

49. Look up more information about this feature. In your own words, describe how it formed and include its age.

50. In which Oregon geological province is this located?

Southeastern Oregon

Instructions

Enter all your responses in the Google Form.

1. Navigate to the southeastern quadrant of Oregon (centered around the Steens Mountain area but stay at an eye alt of 150-300 miles).
2. Carefully investigate the landscape and identify some horsts and grabens.

> Hint: making a few west-to-east profiles may help you see the "ups and downs" (not required, though).

3. Using the "Path" tool, clearly outline two horsts in red and two grabens in blue. Make the weight of these 2.0.
4. Take a screenshot of these and insert it in the Google Form.

51. In which two Oregon geological provinces is this located? Refer to Figure 8.6.

Figure 8.6. A map of the Oregon Geological Provinces.

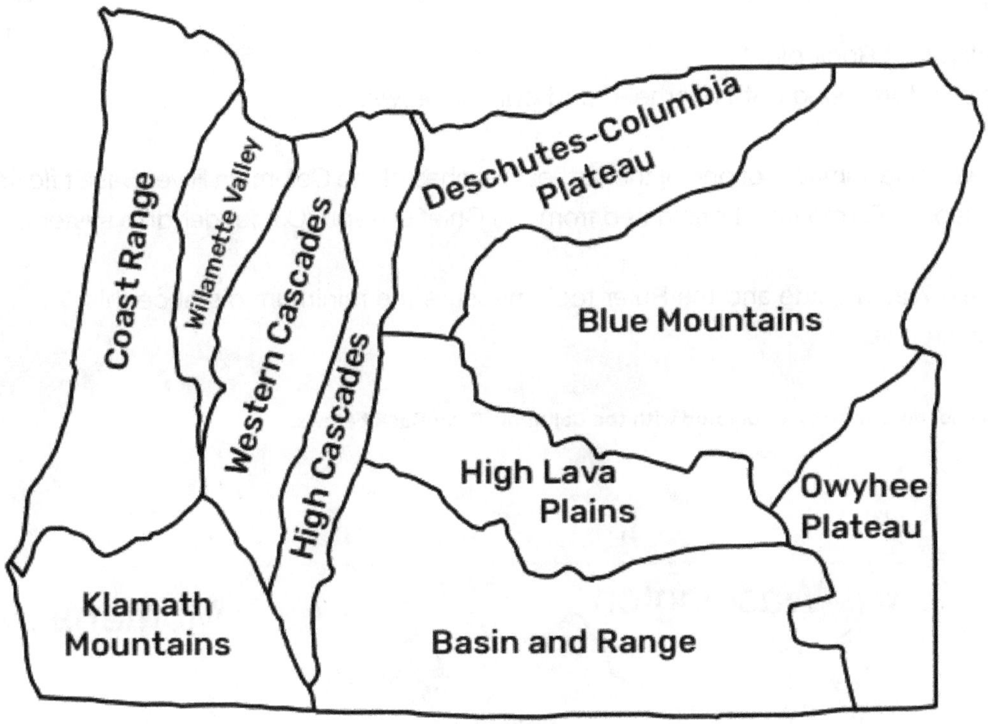

Elephant Rock, Oregon

Instructions

Enter all your responses in the Google Form.

1. Go to the Elephant Rock pin.
2. Observe this feature using both overhead and oblique views.

This feature is an erosional remnant of one of the farthest reaches of the Columbia River Basalt flows. As part of the Wanapum Formation, it originated from the Chief Joseph (CJ) feeder dike system.

52. Using Figure 8.7 as a guide and the Ruler tool, measure the minimum distance this lava traveled (in miles).

Figure 8.7. Primary feeder dike swarms associated with the Columbia River Basalt Flows.

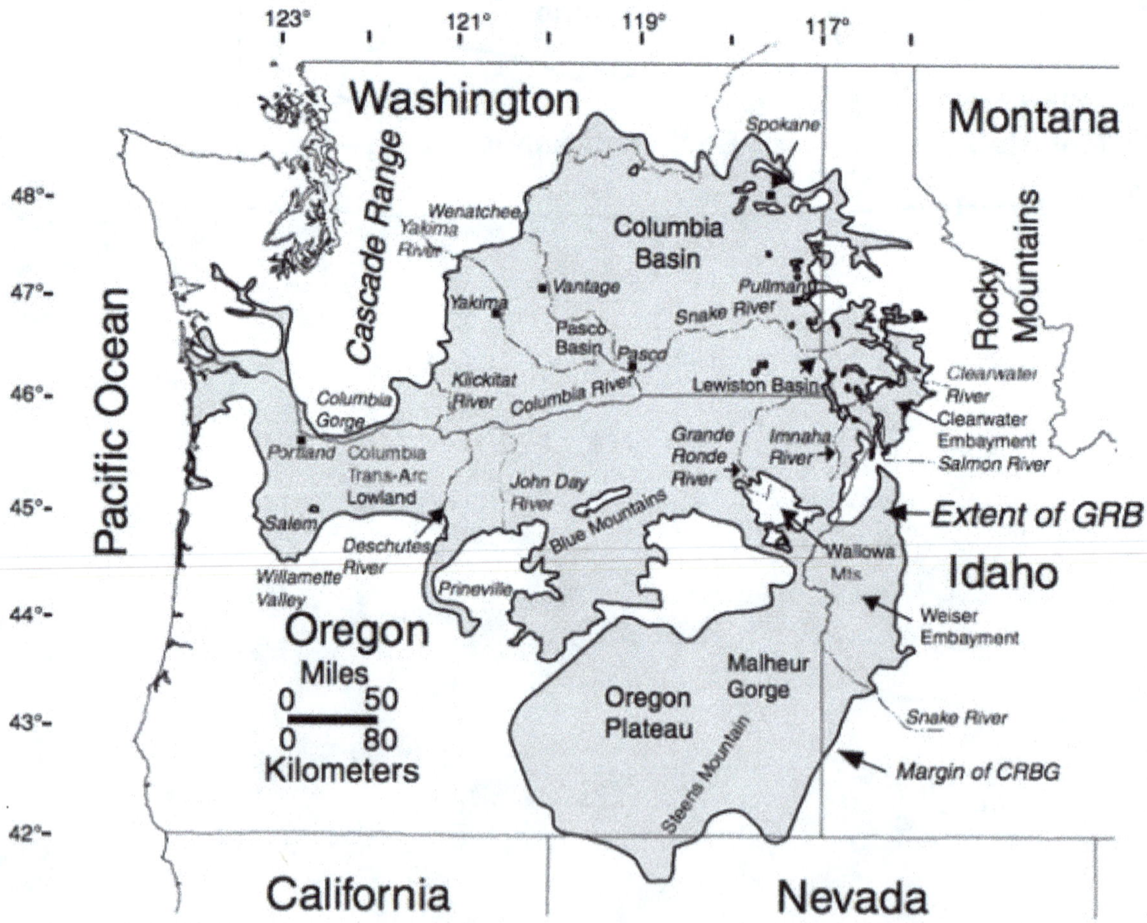

Name: ______________________ Lab Time: ______________ Due: ______________

9 Profiling Mt. Saint Helens

Purpose

In this lab exercise, you will analyze the change in shape, elevation, and gradient of Mount St. Helens after its 1980 eruption. You will also review a PBS video detailing the 1980 eruption, subsequent activity, and its "return to life" from both a geological and biological aspect.

Objectives

- ❑ Construct topographic profiles from contour maps.
- ❑ Compare the topography of Mt St Helens before and after the 1980 eruption.
- ❑ Calculate changes in summit elevation and slope gradient resulting from the eruption.
- ❑ Use a simplified geometric model to estimate the volume of material removed during the eruption, and evaluate the model's limitations.
- ❑ Explain how the 1980 eruption reshaped the volcano and altered volcanic hazards in the surrounding landscape.

Key Terms

- ❑ Topography: the shape and arrangement of Earth's surface features, including changes in elevation and relief.
- ❑ Contour line: a line on a topographic map that connects points of equal elevation on land.
- ❑ Contour interval: the change in elevation between each contour line shown on a map.
- ❑ Topographic profile: a graph showing how elevation changes along a line drawn across a landscape. Vertical exaggeration is sometimes used to make elevation changes easier to see.
- ❑ Transect: a specified line on a topographic map.
- ❑ Relief: difference in elevation between the highest and lowest points in a map area or along a transect (calculated by subtracting low elevation from high elevation).
- ❑ Volume: the amount of space something takes up in 3 dimensions.
- ❑ Gradient: the steepness of a slope, usually calculated as the following equation:

$$\text{Gradient} = \frac{\textit{change in elevation (rise)}}{\textit{horizontal distance (run)}}$$

9

Materials

- ❑ Calculator
- ❑ Pencil with eraser
- ❑ Ruler or straight edge

Introduction

Mount Saint Helens (MSH) is one of 13 composite-cone volcanoes that together form the most prominent peaks of the Cascade Range, which extends from northern California into Canada. MSH, in southern Washington State, is currently the most active member (Figure 9.1).

Figure 9.1. Major Cascade Range Volcanoes. MSH shown in red. (USGS.gov)

The Cascade volcanoes result from the convergence of the Juan de Fuca plate (oceanic crust) and the North American Plate (continental crust), where subduction heats the descending lithosphere, causing partial melting and magma production, resulting in periodic lava flows and pyroclastic eruptions. The lava is typically andesitic or dacitic (between andesite and rhyolite) in composition.

On May 18, 1980, Mount Saint Helens suddenly erupted, killing dozens of people. It was the deadliest and most explosive volcanic event in recorded U.S. history. The explosion was unique in that it was directed laterally (sideways) and northward. The volume of ejecta is estimated at around 0.3 cubic miles (1.25 cubic kilometers), placing it at a VEI of 5 (barely). See Figure 9.2.

Figure 9.2.: Shaded elevation model of MSH before (above) and after (below) the 1980 eruption. (USGS.gov)

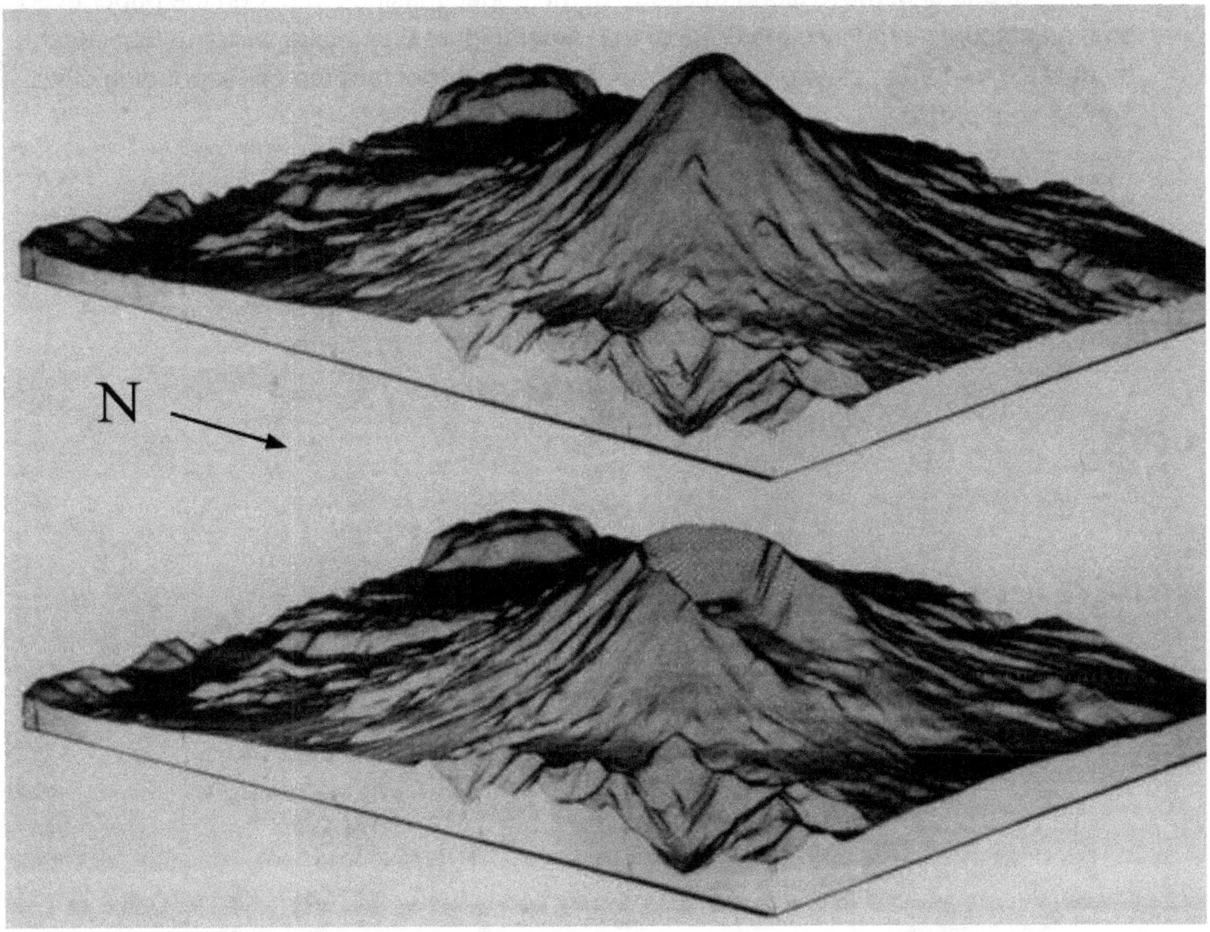

The 1980 eruption dramatically reshaped Mount St. Helens in just a few minutes. By comparing topographic maps from before and after the eruption, we can quantify how volcanic processes alter Earth's surface and better understand why topographic maps are valuable tools for studying geologic hazards.

Part 1. Drawing a Topographic Profile

A topographic profile is essentially a cross-sectional view of a map area along a specific line. Topographic profiles can be helpful because they visually show the relief of an area of interest. Follow the steps below.

Instructions

1. Locate the transect line on the map.
2. Place a blank piece of paper along the line and mark the starting and ending points of the line.
3. Indicate where the transect begins and ends with designated letters.[1]
4. Starting at one end, move along the edge of the paper, making a mark on the paper every time a contour line touches the edge of the paper (Figure 9.3). Make sure you label each mark with the correct elevation so you can transfer that point to the corresponding elevation on your profile.

Figure 9.3. Making a transect line.

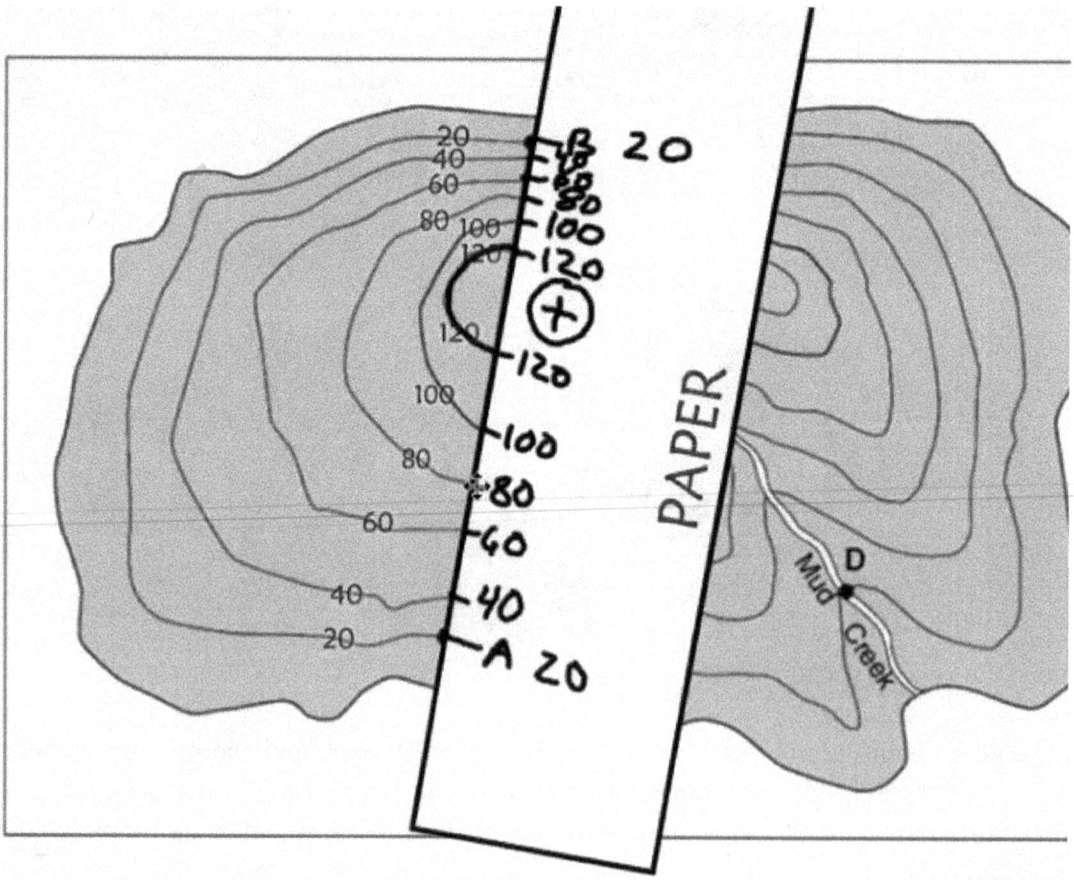

1. Transects are usually shown as A-B ("A to B") or X-Y ("X to Y") where A or X indicate the starting point and B or Y indicate the ending point.

5. For every mark you made on the paper, place a vertical point on the graph directly above that point corresponding to the correct elevation (Figure 9.4). Use the edge of the paper as a straight edge.
6. Connect the dots on the graph paper to complete the topographic profile. Connect the plotted points with a smooth curve that follows the shape of the landscape. Do not draw stair-step lines between contour intervals.

Helpful hints and reminders:

- ❑ Make every contour crossing.
- ❑ Label each tick with its elevation immediately.
- ❑ Use the same horizontal scale for your graph and map (don't change spacing).
- ❑ Plot points directly above their tick marks.

Before continuing, have your instructor verify that your first profile was done correctly!

Figure 9.4. Creating a profile on a graph.

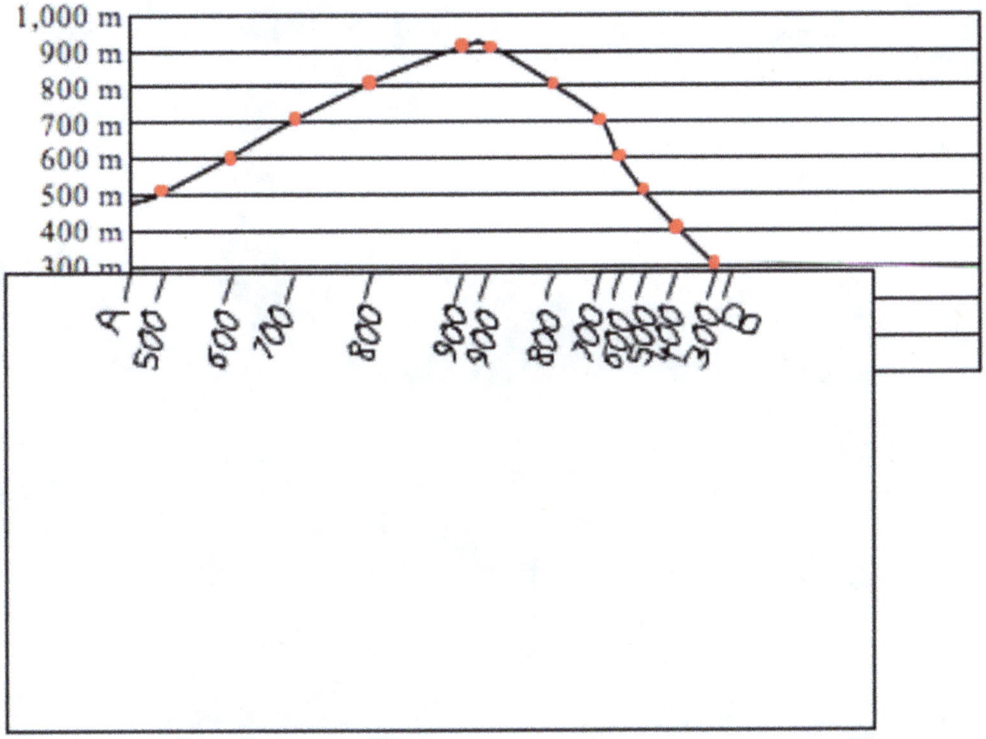

7. Repeat this process for the following profiles:
   a. Using the pre-eruption map of MSH (Figure 9.5), complete an A – B profile (Figure 9.6).
   b. Using the post-eruption map of MSH (Figure 9.7), complete an A – B profile (Figure 9.8).
   c. Using pre-eruption and the post-eruption maps of MSH, complete an X-Y profile for each (Figure 9.9).

Part 2. Initial Observation Questions

Reviewing Volcanic Eruption Maps

Instructions

1. Observe the two topographic maps in Figures 9.5-9.9.

1. Which side of the volcano do you think changed the most during the eruption? What evidence on the maps supports your prediction?

2. Do you expect the volcano to be taller or shorter after the eruption?

3. Which transect (A-B or X-Y) do you predict will show the greatest difference before and after the eruption? Explain your reasoning.

9

Figure 9.5. Mt. St. Helens before the 1980 eruption. (USGS.gov)

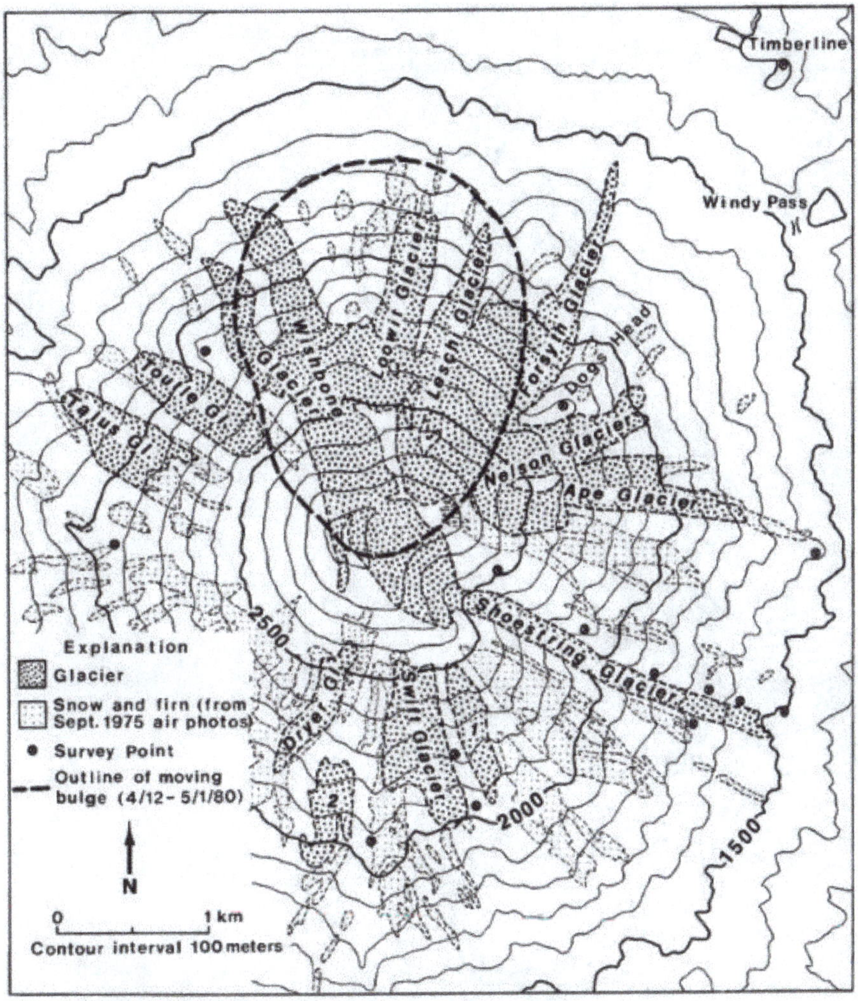

Figure 9.6. Mt. St. Helens pre-eruption Profile (A-B)

Figure 9.7. Mt. St. Helens after 1980 eruption (USGS.gov)

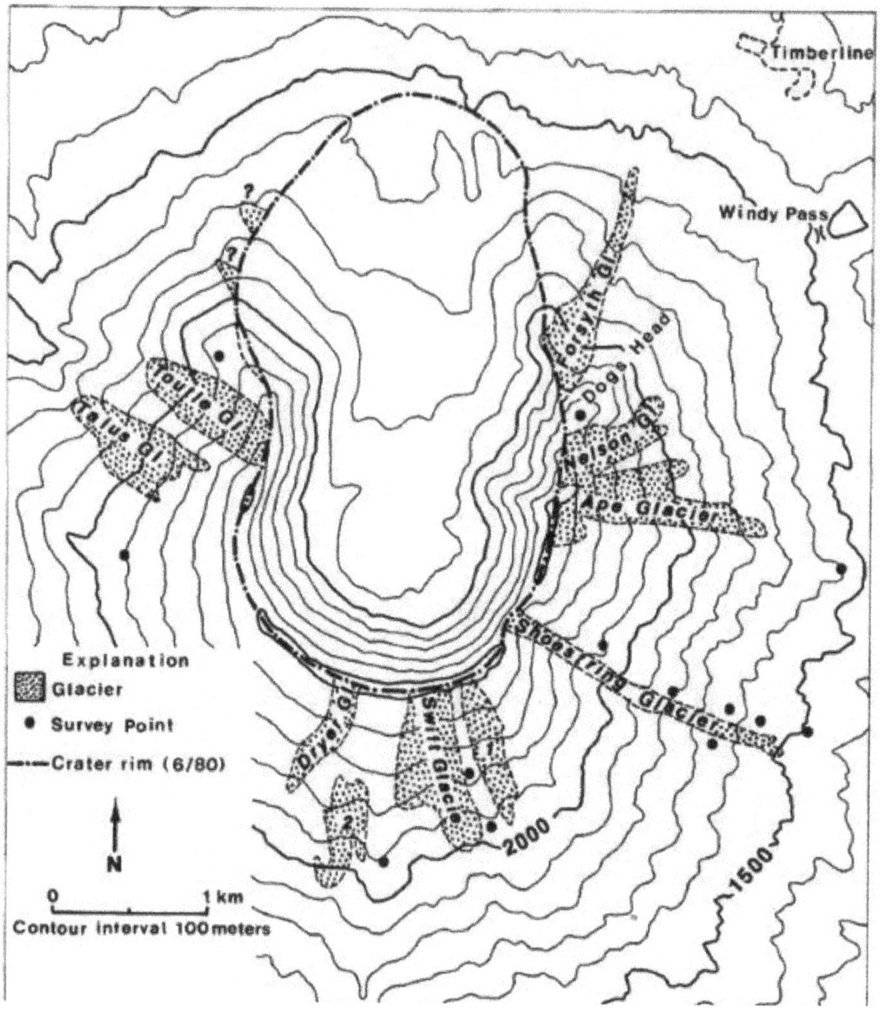

Figure 9.8. Mt. St. Helens post-eruption Profile (A-B)

Figure 9.9. Mt. St. Helens pre- and post-eruption gradient profiles (X-Y). Plot pre-eruption on the left, post-eruption on the right.

4. Identify two major differences between the pre-and post-eruption A-B profiles.

5. Which part of the volcano experienced the greatest amount of change?

6. Based only on the profiles, what evidence suggests that the eruption blast was directed to the north (as opposed to all directions equally)? Be specific.

7. Which profile best illustrates why the 1980 eruption is described as a "lateral blast" rather than a typical vertical eruption?

Calculating Eruption Changes

8. What was the summit elevation before the eruption?

9. What was the summit elevation after the eruption?

Instructions

9

1. Calculate the elevation change.[2]

10. Compare the pre- and post-eruption profiles again, and now consider volume loss. Estimate which of the following best shows the approximate volume loss AND explain your reasoning using your profiles.
    a. <5%
    b. 5-10%
    c. 10-20%
    d. >20%

Volume loss caused by explosive eruptions of composite cones (stratovolcanoes) is often estimated using a simplified cone model where the missing portion (volume) of the volcano is approximated to be a cone. While MSH is not a perfect cone, Figure 9.10 provides a reasonable first-order estimate of the volume of rock removed by the eruption. It's modeled as a cone using the following

2. Fun fact: Prior to the 1980 eruption, MSH was the 5th highest peak in Washington State. After the eruption (and with ongoing erosion), it is now ranked 102nd!

formula, where r = the radius of the crater and h = the elevation lost (pre-eruption summit elevation- post -eruption summit elevation).

Measurement hints:

- ❑ use the post-eruption map to measure the approximate radius of the missing summit (crater).
- ❑ record both measurements in km.
- ❑ measure to the nearest 0.1 km.
- ❑ radius is the straight-line distance from the center of a circle to any point on its outer edge.

Figure 9.10. A first-order estimate of the volume of rock removed by the eruption.

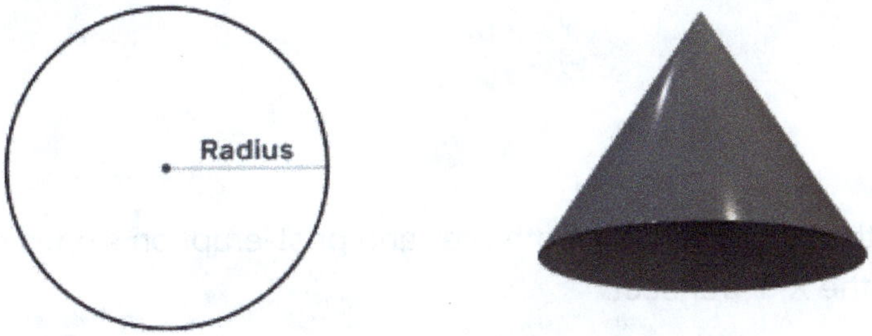

$$V = \frac{1}{3}\pi r^2 h$$

2. Complete Table 9.1 and use the equation above to calculate volume loss (V).

Table 9.1. Calculate the Volume Loss

	r	h	V
3.14	________ km	________ km	________km3

11. Scientists estimate that approximately 2.5 km² of the volcano collapsed during the 1980 eruption. Compare this value with your estimate. Is your estimate larger or smaller? Explain at least two reasons why your simplified model estimate differs from the published value. Be specific.

12. How can volume loss be determined more accurately?

13. What volcanic process or processes caused the summit to lose elevation and volume during the 1980 eruption?

3. Complete Table 9.2 to calculate the % gradient for both pre- and post-eruption slopes on the northern flank marked along the X-Y transect.

Table 9.2. Calculate the % Gradient for the Pre- and Post-eruption Slopes

	Highest elevation/ top (m)	Lowest elevation/ bottom (m)	Elevation change (m)	Horizontal distance (m)	Gradient (m/m)	Gradient (%)
1. Pre-eruption						
2. Post-eruption						

14. Based on your calculations, which slope (before or after) is steeper?

15. How might this change influence future hazards such as landslides or lahars? Consider slope stability, water flow, and loose volcanic material in your answer.

16. Which profile better represents a typical composite cone (stratovolcano)? Explain.

17. Assume a pause in volcanic activity on MSH for several hundred years. Predict how erosion and deposition would change the A-B profile over time.

Part 3. MSH: Back from the Dead (video)

- ❑ How many people died because of the eruption?
- ❑ How deep is the crater?
- ❑ What surprised scientists about the return of life to Mount St. Helens? Give one specific example discussed in the video.
- ❑ Beginning in 2004, Mount St. Helens entered a new eruptive phase. How did this eruption differ from the 1980 eruption?
- ❑ How did the microscopic investigation of extrusive rock produced from the 2004–2008 event explain its lack of explosivity?
- ❑ How did the 1980 eruption allow us to "see into the past" in order to understand previous eruptions?
- ❑ Imagine that the 1980 eruption occurred today under the same conditions. Describe three hazards nearby communities would face. Use evidence from both your profiles and the video in your response.

9

Name: ______________________ Lab Time: ______________ Due: ______________

10 Earthquakes

Purpose

Earthquakes provide valuable information about Earth's interior and the movement of tectonic plates. In this four-part investigation, you will analyze earthquake data to answer a series of questions about earthquakes and plate tectonics.

Objectives

- ❑ Locate an earthquake epicenter using seismic-wave arrival times and triangulation.
- ❑ Determine earthquake magnitude using seismic-wave amplitudes and a Richter nomogram.
- ❑ Calculate and interpret earthquake recurrence intervals using historical earthquake records.
- ❑ Interpret global earthquake and volcanic distributions to identify major plate boundary types.
- ❑ Interpret earthquake hypocenter data to identify a Wadati-Benioff zone.

Key Terms

- ❑ **Seismology**: the study of earthquakes and seismic waves.
- ❑ **Hypocenter**: the point beneath the Earth's surface where an earthquake begins.
- ❑ **Epicenter**: the point on Earth directly above an earthquake hypocenter.
- ❑ **Epicentral distance**: the distance between a seismic station and the epicenter of a recorded earthquake.
- ❑ **Seismic wave**: shock waves generated by an earthquake that travel outwards in all directions from the hypocenter.
- ❑ **S-P Interval**: the time delay between the detection of the first P wave and the first S wave generated from an earthquake.
- ❑ **Seismograph**: a device that measures vibrations of the ground.
- ❑ **Seismogram**: a graph produced by a seismograph showing ground motion as a function of time.
- ❑ **Nomogram**: a graphical chart used to quickly find the Richter magnitude of an earthquake based on epicentral distance and maximum S-wave amplitude.
- ❑ **Subduction**: a place on Earth where two tectonic plates converge and one of them is forced beneath the other.
- ❑ **Wadati-Benioff Zone**: the zone of earthquakes that descends into the Earth on a subducting plate.

Materials

- ❑ Calculator
- ❑ Colored pencils (red, blue, green)
- ❑ Compass (or string)
- ❑ Pencil with eraser
- ❑ Ruler or straight edge

Part 1. Locating an Earthquake Epicenter and Determining Richter Magnitude

Introduction

Here we will look at three seismograms recorded at three different seismic stations across the Pacific Northwest to locate the epicenter and determine the Richter Magnitude for a real historical earthquake.

Because seismic waves travel outward in all directions from an earthquake, every seismic station records the earthquake but at different times. The difference between the arrival times of the P and S waves depends on the station's distance from the epicenter. By determining this distance for three or more stations and drawing circles of equal radius around each station, the earthquake epicenter can be located at the point where the circles intersect.

Determining the Richter Magnitude of Earthquakes

Instructions

1. Observe the three seismograms given in Figure 10.1. Note that the X-axis shows time (in seconds) and the Y-axis shows seismic wave amplitude (in millimeters).

Figure 10.1. Seismograms of Boise, Portland, and Eureka.

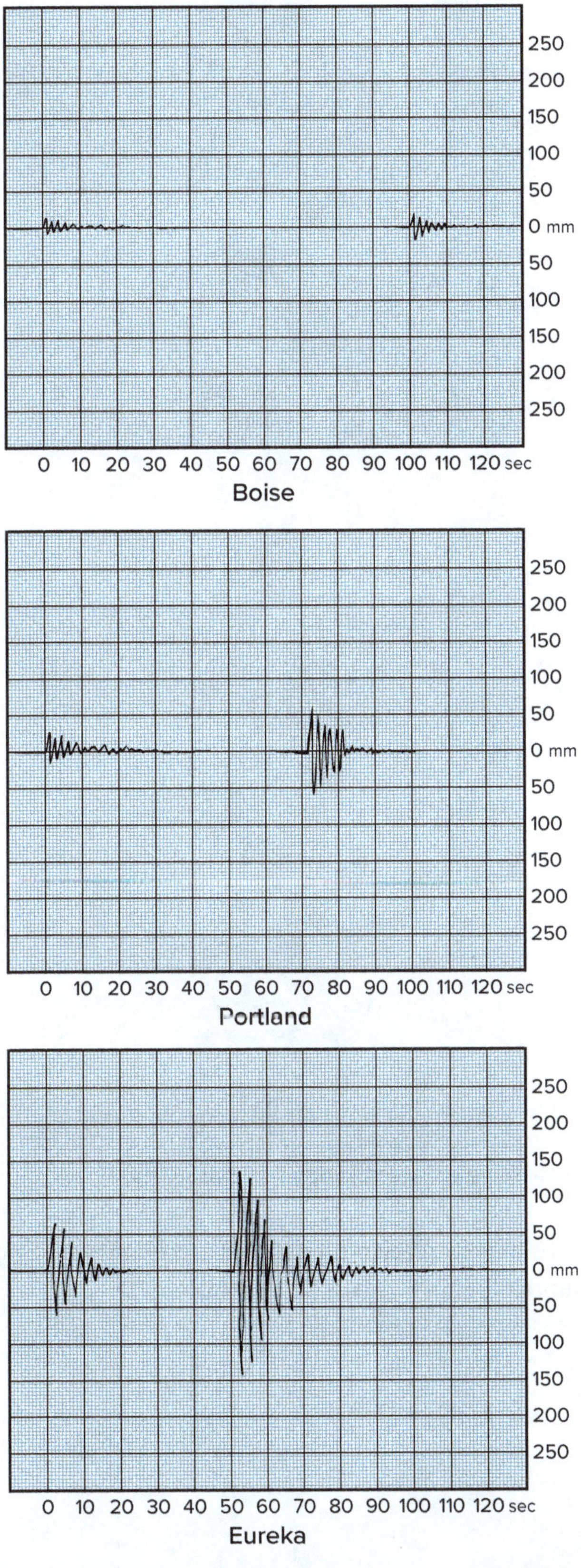

2. Observe the geographic locations of each corresponding seismic station on the map shown in Figure 10.2.

Figure 10.2. Map of western US with seismic stations indicated by triangles.

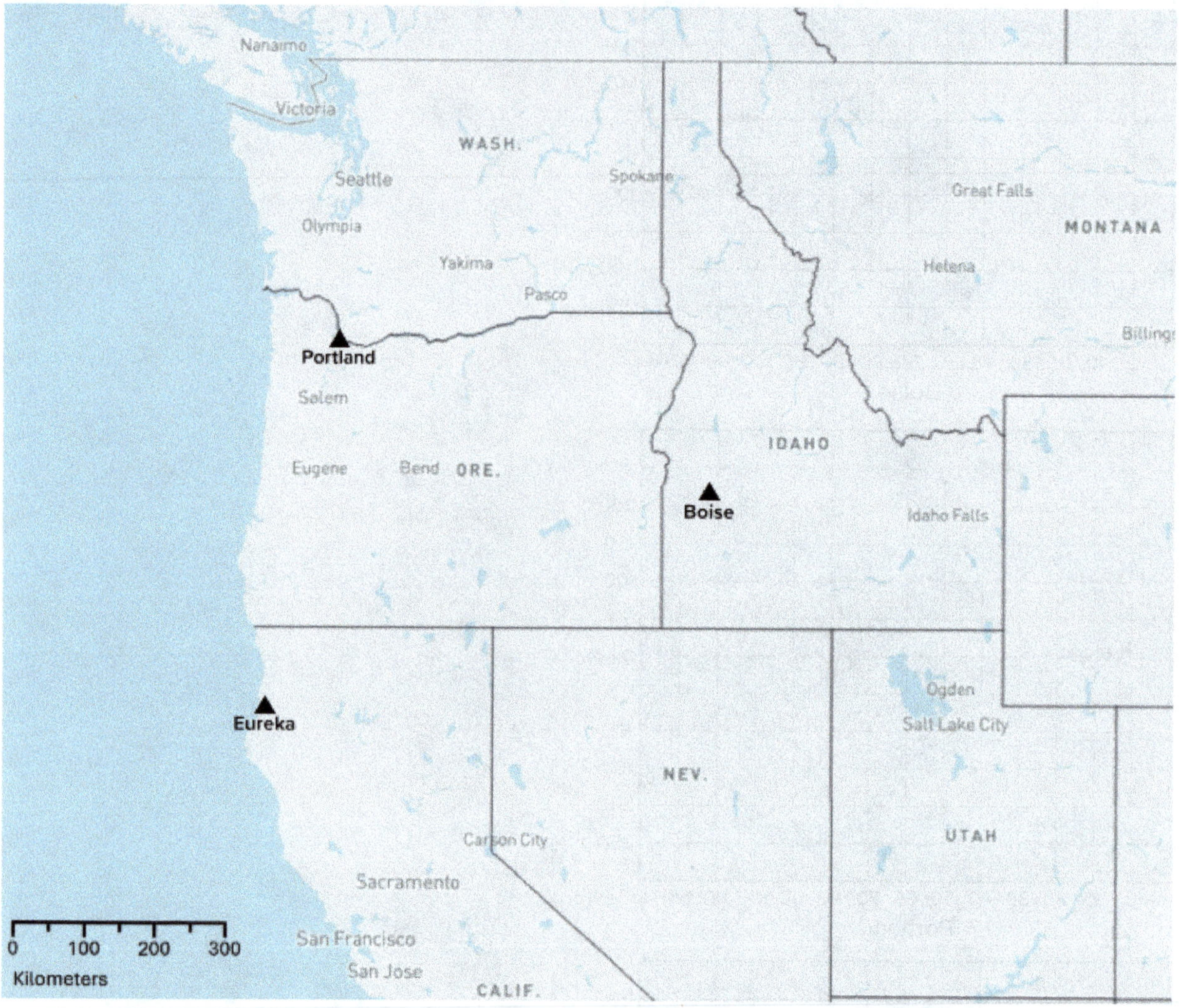

3. On each seismogram, label the P-wave and S-wave clusters (above or below the line).

1. Compare the overall amplitudes of the S-wave cluster shown on each seismogram. List the seismograms according to overall amplitudes. (1=greatest, 3=smallest)

____

____

____

2. Compare time delay of the P and S wave clusters for each seismogram. List the seismograms according to delay between P and S waves. (1=closest together, 3=farthest apart)

____

____

____

3. Based on your rankings, which seismic station do you predict is closest to the earthquake epicenter? Which is farthest away? Explain your reasoning.

4. On each seismogram, indicate where each set of waves starts. Place a circle where each set is first detected.
5. Determine the S-wave arrival time for each station and record it in Table 10.1. Because the P-wave arrival has been shifted to 0 seconds, the S-wave arrival time is equal to the S–P interval.
6. Using the S-P interval calculated for each station and the P and S Wave Time Travel Graph (Figure 10.3), determine the distance to the epicenter. For each station:
   a. Locate the S-P interval on the y-axis.
   b. Use a piece of scrap paper to mark this vertical distance.
   c. Slide the paper horizontally until the spacing between the P and S curves matches your mark.
   d. Read the corresponding distance from the x-axis.
   e. Record this value in Table 10.1.

Example: an S-P interval of 170 seconds corresponds to an epicentral distance of about 900 km.

10

Figure 10.3. P and S Wave Time Travel Graph. This graph is intentionally simplified to illustrate the relationship between S-P interval and epicentral distance. It is not intended to represent actual seismic-wave velocities within Earth.

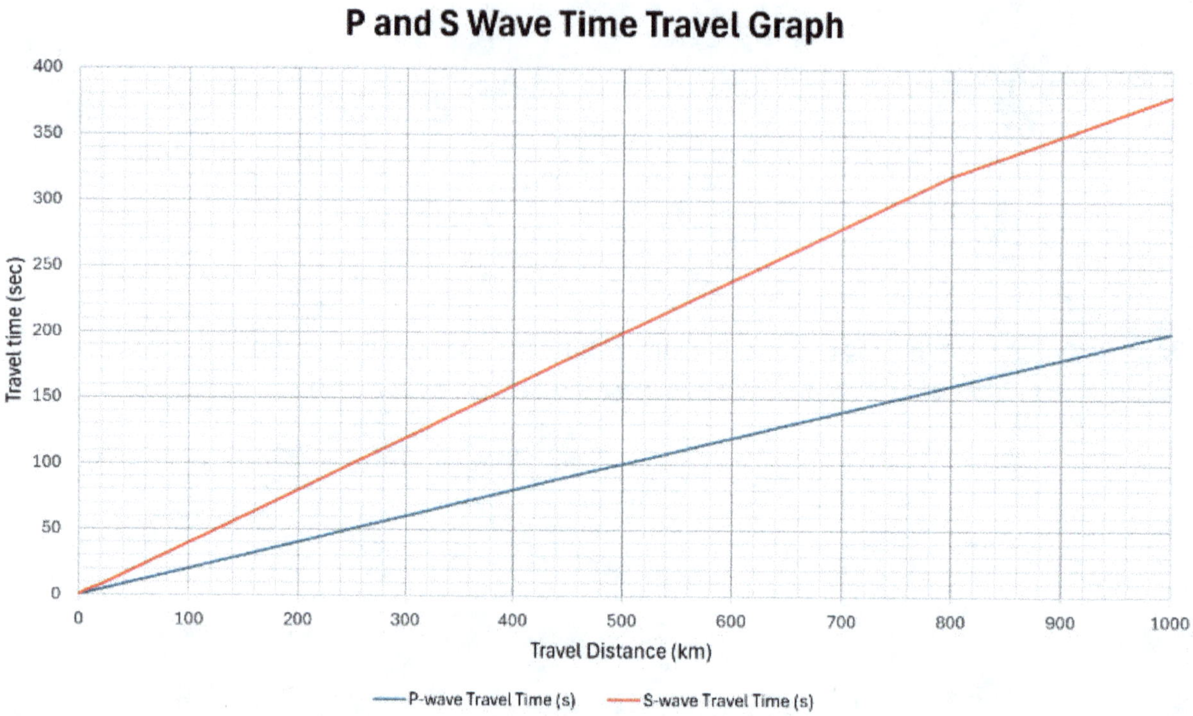

Table 10.1. Calculating Epicentral Distance

Seismic station	P-wave arrival time (sec)	S-wave arrival time (sec)	S-P Interval (sec)	Epicentral distance (km)
Boise, ID				
Eureka, CA				
Portland, OR				

7. Triangulate to find the epicenter. For each station:
   - ❑ Using the bar scale on the map (Figure 10.2), open your compass to the corresponding epicentral distance calculated above. Measure carefully.
   - ❑ Keeping the compass steady (don't let the angle change), place the point directly on the seismic station ( ) and draw an arc on the map that maintains this distance. You do not need to continue the arc beyond the boundaries of the map.
   - ❑ Trace each arc with colored pencils: green for Boise, red for Eureka, and blue for Portland.
   - ❑ The location where all 3 arcs intersect is the epicenter. Mark the epicenter with a prominent yellow dot.

Note: If your arcs do not intersect, first verify your S-wave arrival times before checking your compass measurements. Because of measurement uncertainty, your three circles may not intersect at exactly one point. Estimate the epicenter as the center of the small region where the circles overlap most closely.

8. Determine the Richter Magnitude for this earthquake using the nomogram (Figure 10.4). For each station:
   a. Record epicentral distance values in Table 10.2. (These are the same values calculated and found in Table 10.1.)
   b. Determine the maximum S-wave amplitude (in millimeters) for each seismogram and record it in Table 10.2.
   c. For each station,
   d. Mark the epicentral distance value on the Distance axis with a dot.
   e. Mark the maximum S-wave amplitude value on the Amplitude axis with a dot.
   f. Draw a straight line connecting corresponding dots.
   g. Trace each line with colored pencils: green for Boise, red for Eureka, and blue for Portland.
   h. All three lines should cross the middle axis (Magnitude) at the same place. This is the Richter Magnitude value. Record in Table 10.2.

Figure 10.4. Nomogram for Richter magnitude calculation. Epicentral distance (left axis) and maximum S-wave amplitude (right axis) are needed to determine Richter scale magnitude (center axis).

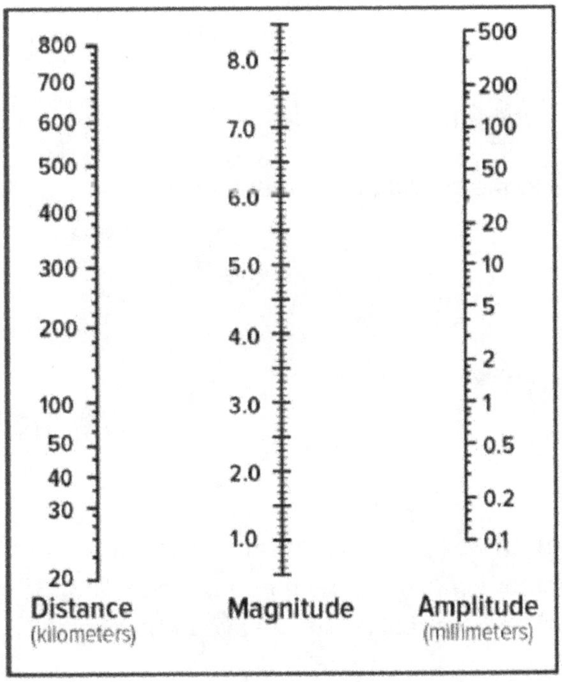

10

Table 10.2. Calculating Richter Magnitude

	Epicentral Distance (km)	Maximum S-wave amplitude (mm)	Average Richter Magnitude value:
Boise, ID			
Eureka, CA			
Portland, OR			

4. Which of the following best fits the location of the earthquake epicenter?
   a. between Salt Lake City and Ogden, UT
   b. in the Pacific Ocean, southwest of San Francisco, CA
   c. between Sacramento and San Francisco, CA
   d. southern Oregon
   e. western Montana
   f. way off the map…I think I did something wrong

5. Which of the following cities should receive the seismic waves first?
   a. Bend, OR
   b. Sacramento, CA
   c. Ogden, UT
   d. Helena, MT
   e. Idaho Falls, ID
   f. Seattle, WA

10

6. Estimate the distance from Carson City, Nevada to the earthquake epicenter using the completed map.
   a. 120 km
   b. 250 km
   c. 400 km
   d. 725 km
   e. 1150 km

7. The P and S Wave Time Travel Graph (Figure 10.3) was simplified for this introductory geology course. Describe two assumptions or simplifications made by the P and S Wave Time Travel Graph. Explain how each differs from real seismic wave behavior.

Part 2. Cascadia Recurrence Interval

Introduction

Although earthquakes cannot be predicted, geologists can estimate how frequently earthquakes of different magnitudes occur over long periods of time. These recurrence intervals help scientists estimate the probability of future earthquakes and evaluate earthquake hazards.

The Cascadia Subduction Zone is capable of producing earthquakes larger than any historically recorded in the Pacific Northwest. This would pose a serious threat to coastal communities and large urban areas like Portland, Seattle, and Vancouver, B.C. But is it possible to determine how likely it is that a large earthquake will strike the Pacific NW? Seismologists use conditional probabilities (the likelihood that a specific event will occur, given that another event has already occurred) to estimate the probability that an earthquake of a given magnitude will occur in an area within a certain period of time.

In this activity, you will use historical earthquake records to calculate recurrence intervals for earthquakes of different magnitudes and use these trends to estimate how often great Cascadia earthquakes may occur. These estimates are based on real data collected from the US Geological Survey (USGS) and historical records from the Pacific Northwest Seismic Network (PNSN). In this activity, we will analyze the seismic history of earthquakes in the Pacific Northwest to estimate the hazards we might be facing in the future.

Calculating Cascadia Recurrence Intervals

Using the data given on the following pages, complete the tables and answer the questions.

Instructions

1. Observe the data in Table 10.3. Note that three smaller tables show the number of earthquake events within a given magnitude range over a given time period.
2. Determine the total number of earthquake events for each data set and record it in Table 10.3.

Table 10.3.A-D. Earthquake Events and Magnitude

10.3.A. Magnitude 2.0-2.9 Earthquakes

Two-year period	Number of M 2.0-2.9 earthquakes
1992-1993	113
1994-1995	62
1996-1997	59
1998-1999	57
2000-2001	49
Total in ten-year period:	

10.3.B. Magnitude 3.0-3.9 Earthquakes

Three-year period	Number of M 3.0-3.9 earthquakes
1988-1990	52
1991-1993	151
1994-1996	84
1997-2000	108
2001-2003	68
Total in 15-year period:	

10.3.C. Magnitude 4.0-4.9 Earthquakes

Five-year period	Number of M 4.0-4.9 earthquakes
1979-1983	151
1984-1988	39
1989-1993	45
1994-1998	31
1999-2003	29
Total in 25-year period:	

10.3.D. Magnitude 5.0-7.9 Earthquakes

Magnitude	Time range	# of years	# of earthquakes
5.0-5.9	1973 - 2003	30	25
6.0-6.9	1872 - 2003	131	9
7.0-7.9	1872 - 2003	131	3

10

Complete Table 10.4 by filling in Columns 1-7.

a. In Column 2, indicate the number of earthquakes of each magnitude range that you previously totaled.

b. Determine the average number of earthquakes in 1 year (Column 4): divide the total number of earthquakes for each magnitude by the number of years in your data set.

Calculation: Column 2 / Column3

c. Determine the expected number in ten years (Column 5): multiply the average number of earthquakes expected in one year by 10.

Calculation: Column 4 x 10

d. Determine the time between earthquakes in days (Column 6): divide 365 days per year by the average number of earthquakes in one year.

Calculation: 365 / Column 4

e. Determine the time between earthquakes in years (Column 7): divide the time between earthquakes in days by 365 days.

Calculation: Column 6 / 365

Table 10.4. Recurrence Interval Calculations

1. Magnitude range	2. Number of earthquakes (calculated above)	3. Years in data set	4. Average number in one year	5. Expected number in ten years	6. Time between earthquakes (days)	7. Time between earthquakes (years)
			Calculated Values			
2.0-2.9		10				
3.0-3.9		15				
4.0-4.9		25				
5.0-5.9		30				
6.0-6.9		131				
7.0-7.9		131				

3. Plot the magnitude versus time between earthquakes in years (Column 7 of Table 10.4) on the logarithmic graph paper provided (Figure 10.5).
   a. Plot one point for each magnitude range.

> Note: Plot all magnitude 2.0-2.9 earthquakes as a point on the vertical line labeled 2, all magnitude 3.0-3.9 earthquakes as a point on the vertical line labeled 3, etc.

   b. Do not connect the points until all have been plotted.
   c. Draw one straight best-fit line through the points.

Figure 10.5. Logarithmic graph paper for plotting.

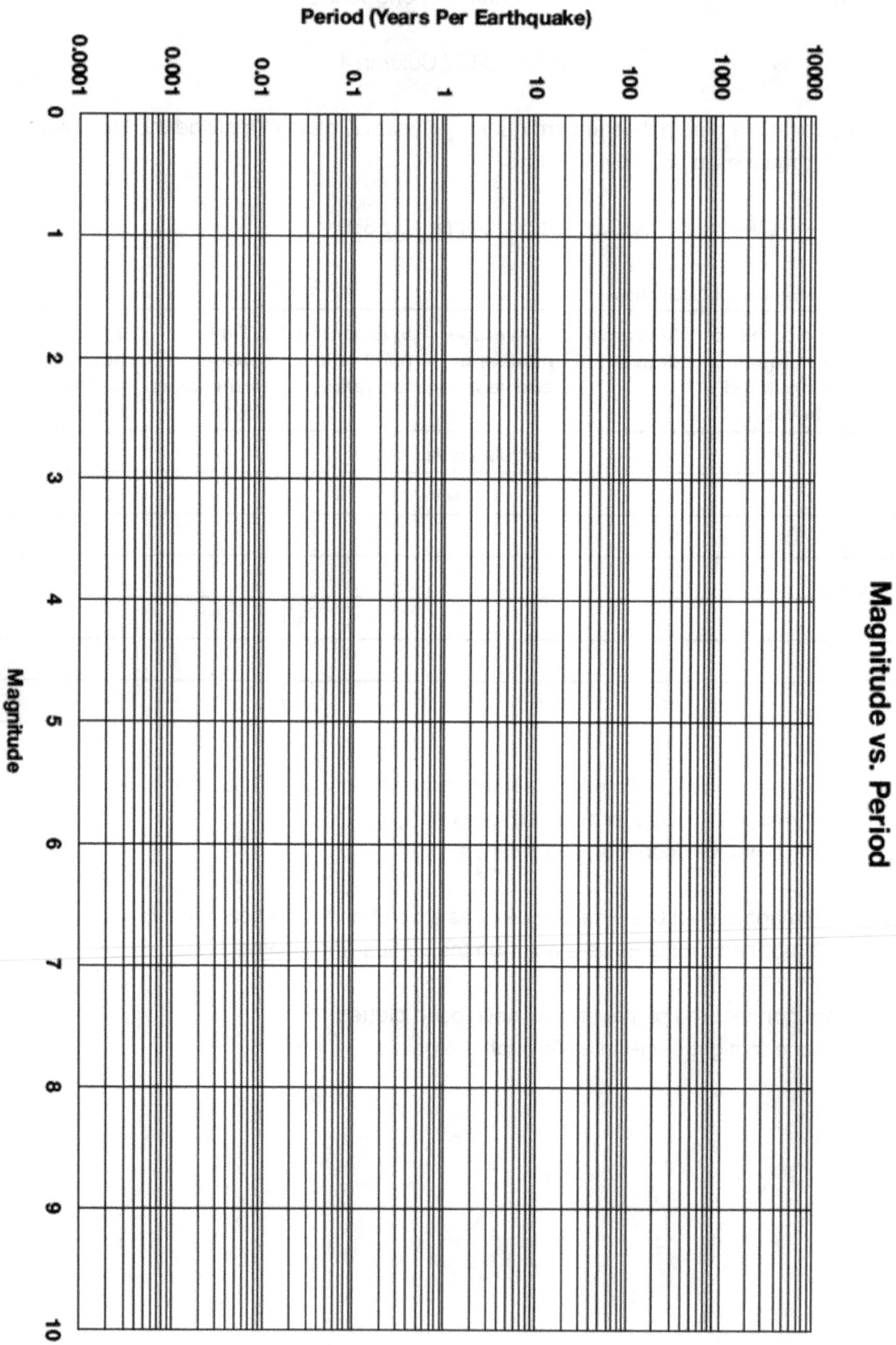

8. Which magnitude range do you think occurs most frequently?

9. Which magnitude range do you think occurs least frequently?

10. How do you think recurrence interval changes with magnitude?

11. What limitations or potential sources of error do you notice in these data? Consider differences in the length of each dataset, unusually active years, and the use of averages.

12. Look at your M 4.0-4.9 data table. Between March 24 and May 19, 1980, there were over 100 earthquakes between M 4.0-4.9. What geologic event caused this increase?

10

13. Should you include 1980 data in your calculations or not? Why or why not? (There is no single correct answer.) Fully discuss your answer.

14. In Table 10.4, observe the values in Column 7 (time between earthquakes in years). What is the relationship between earthquake magnitude and frequency? Explain.

15. Project how often a M 8-9 earthquake might happen by extending a best-fit line through magnitude 8 and 9 on your graph. What is the average recurrence range of earthquakes of this magnitude?

16. Geologic evidence suggests that an M 8-9 earthquake occurred about 325 years ago and 1100 years ago. Do these events fit within your calculations? Explain.

17. Given the information above, discuss the following question: How should emergency planners balance preparation for frequent moderate earthquakes and rare but catastrophic Cascadia earthquakes? Explain your reasoning in detail. Please write at least one paragraph.

18. If the average recurrence interval for M 8–9 earthquakes is 500 years, does that mean another earthquake will occur exactly 500 years after the last one? Explain.

Part 3. Earthquakes and Plate Boundaries

Introduction

Earthquakes are not randomly distributed across Earth's surface. Instead, they occur in distinct patterns that reflect plate motions and interactions. In this part, we will compare global maps of earthquake epicenters and volcanic activity to identify relationships between earthquakes, volcanoes, tectonic plate boundaries, and earthquake depth. You will use these observations to infer the tectonic processes responsible for the observed earthquake and volcanic patterns.

10

Determining Plate Boundaries

Using the maps showing the distribution of global earthquake and volcanic activity, determine the plate boundaries and answer the questions.

Instructions

1. Observe Figures 10.6 and 10.7 on the following pages.

Figure 10.6 shows the global distribution of earthquakes with magnitudes 4.0 and higher during the 6-year period 1990-1996. Each dot represents a single earthquake epicenter. The colors indicate depth to the hypocenter (Red 0-33 km, Orange 33-70 km, Green 70-300 km, Blue 300-700 km)

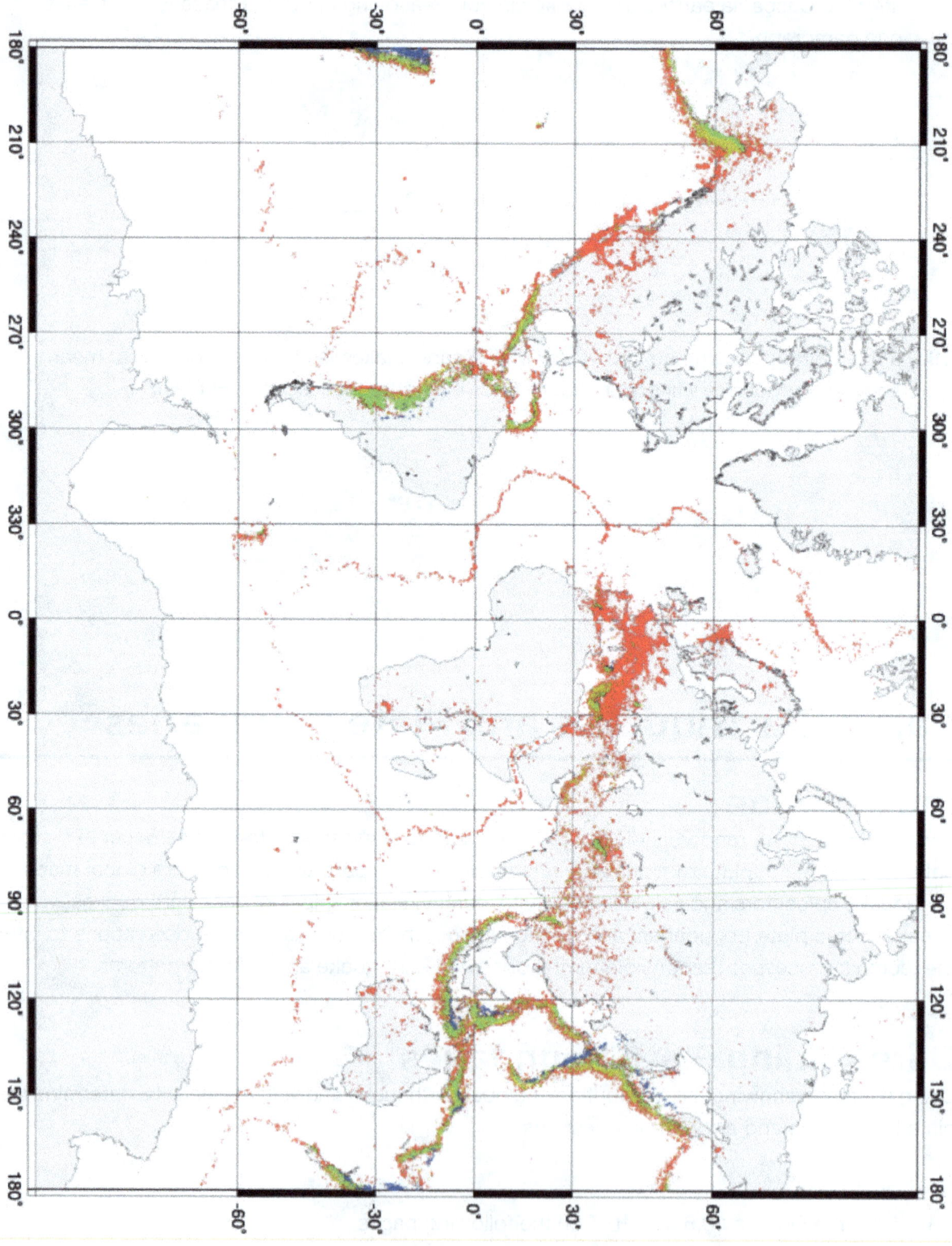

Figure 10.7 shows the global distribution of volcanic activity (currently or historically active).

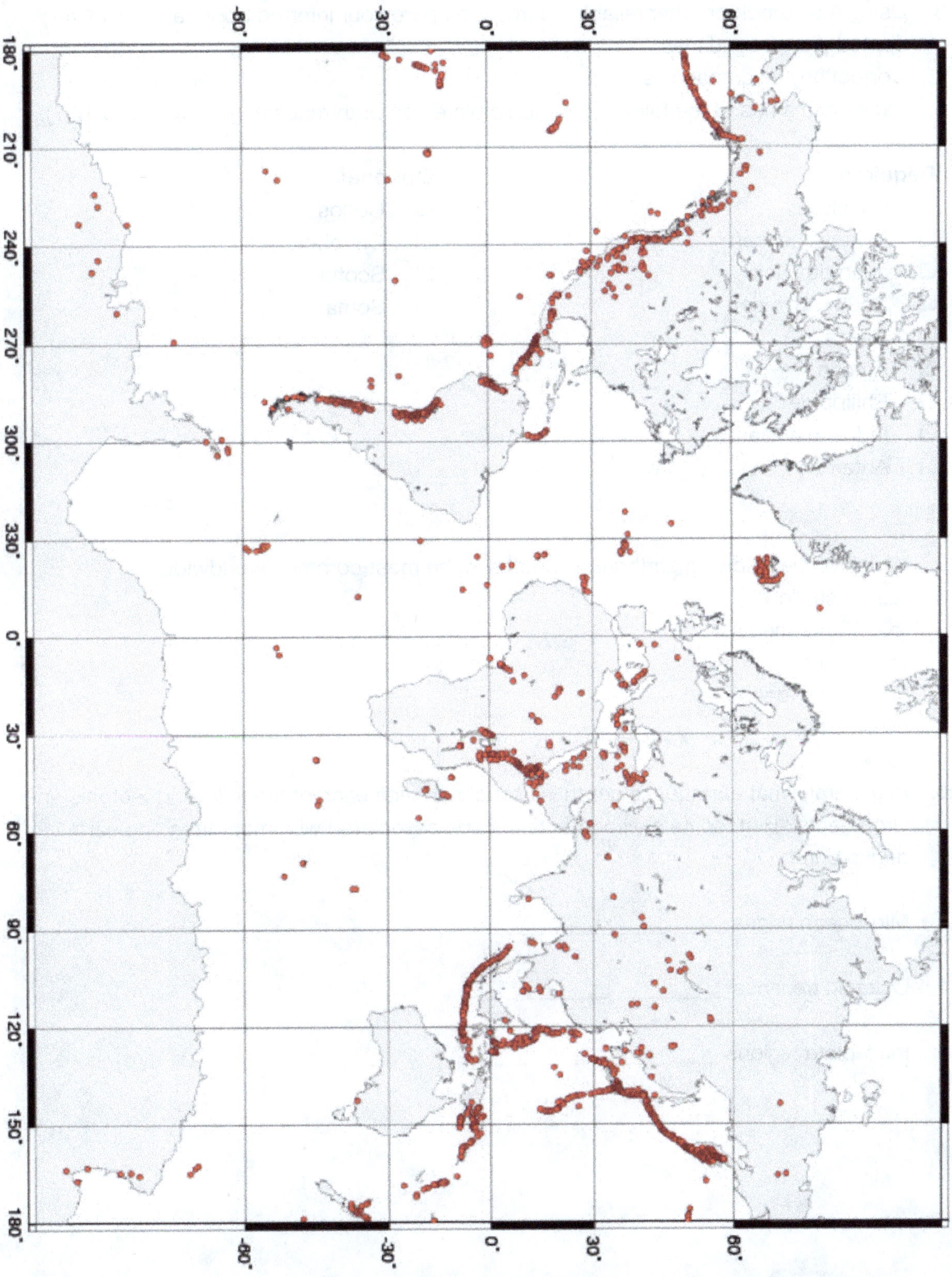

2. On both maps (Figures 10.6 and 10.7), use a pencil to lightly sketch where you think major plate boundaries occur based only on the earthquake and volcano distributions.
3. Using a textbook or other reliable source, compare your inferred boundaries with accepted plate boundaries.
4. Correct any discrepancies.
5. Label the names of the following tectonic plates on both maps (Figure 10.6 and 10.7).

Required

- ❑ Pacific
- ❑ North American
- ❑ Juan de Fuca
- ❑ South American
- ❑ Nazca
- ❑ African
- ❑ Philippine
- ❑ Indo-Australian
- ❑ Antarctic

Optional

- ❑ Cocos
- ❑ Caribbean
- ❑ Scotia
- ❑ Somali

19. Which of the following earthquake depths is the most common worldwide?

a. shallow
b. intermediate
c. deep
d. very deep

20. In general, what earthquake depth is associated with each of the following tectonic environments? Hint: some environments may be associated with more than one earthquake depth.

Mid-ocean ridges ____________________

Oceanic trenches ____________________

Intra-plate regions ____________________

10

21. Indicate the geologic process(es) at each of the tectonic environments (MORs, trenches, intra-plate volcanism) that produce earthquakes of characteristic depths listed above? Explain each.

a. Mid-ocean ridges
b. Oceanic trenches
c. Intra-plate regions

22. Which map (earthquake or volcano) better aligns with tectonic plate boundaries?

Look closely at the Pacific Northwest region (eastern boundary of the Juan de Fuca plate) and answer the following questions.

23. What is atypical about this tectonic environment and the depth of earthquakes observed?

24. Why do you think that atypical seismic behavior is observed here?

25. Based on the two maps and your interpretations, complete Table 10.6. For each location, list at least two map observations (related to volcanoes and seismicity), an inference of the major process(es) occurring there, and an interpretation (the plate boundary type). The first row has been completed for you.

Table 10.6. Earthquakes Across the Globe

Location	Observations	Inference (processes)	Interpretation (plate boundary type)
Japan	earthquakes for a narrow belt depths increase inland volcanoes form a parallel arc	An oceanic plate is descending beneath another plate	Convergent (subduction) boundary
Iceland			
San Andreas			
Hawaii			
Aleutian Islands			
Himalayas			

10

Part 4. Investigating the Wadati-Benioff Zone

Introduction

In subduction zones, earthquakes occur over a wide range of depths. When these earthquake locations are viewed in cross section, they reveal a distinctive pattern that helps geologists understand the movement of tectonic plates beneath Earth's surface. In this activity, you will use real earthquake data from the Tonga subduction zone to determine that pattern.

The Tonga Trench (Figure 10.8) is the second deepest trench in the world, after the Mariana Trench, with a maximum depth of 35,702 ft (10,882 m). It is located in the South Pacific Ocean at the convergent boundary between the Indo-Australian and Pacific tectonic plates.

Figure 10.8. Satellite images of Tonga Trench area. 10a. Regional view of South Pacific area. The red box highlights the map area in 10b. 10b. Tonga Trench near the Kingdom of Tonga. Note the 22°S latitude line. The red triangle represents the andesite volcano referenced in this activity.

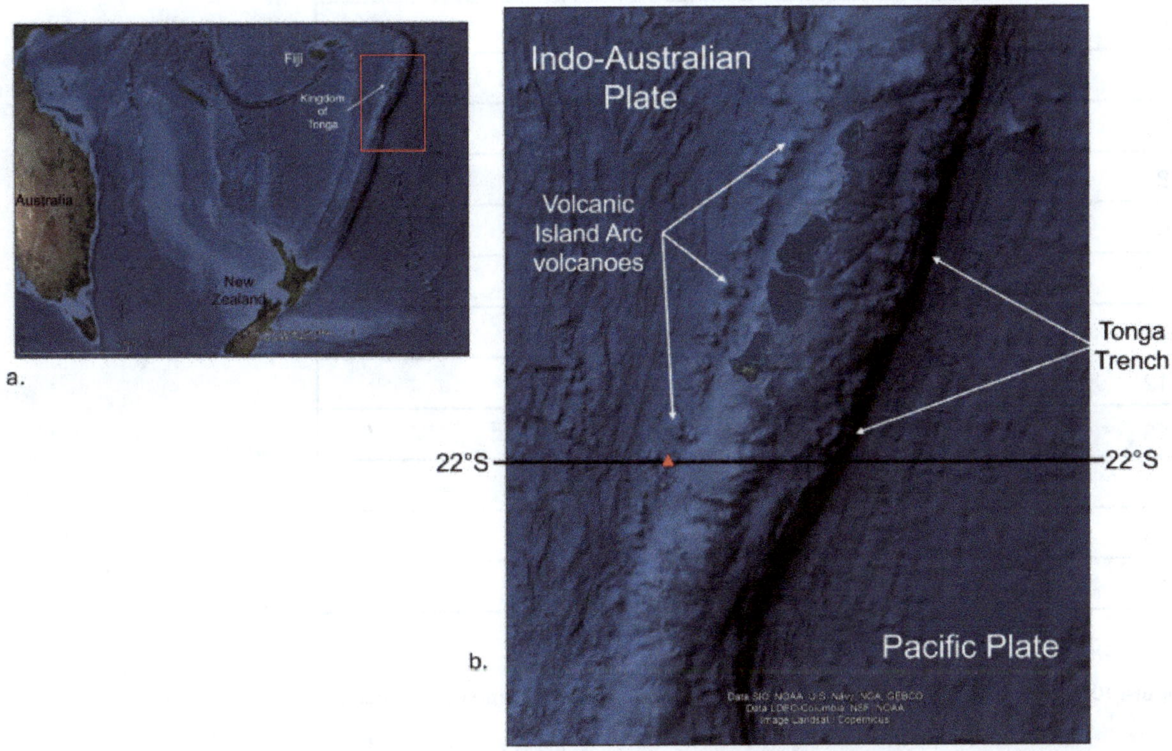

Investigating the Wadati-Benioff Zone

Instructions

1. Observe the location of the Tonga Trench in Figure 10.8. On both the Earthquake and Volcanoes maps from Part III, find and highlight the approximate location of this site.
2. Using the values in Table 10.7, plot the earthquake depths in Figure 10.9. Use a dot or small X for each point.
3. Draw one smooth line through the center of the earthquake cluster to represent the subducting slab. Label this feature the Wadati-Benioff Zone.
4. Label the Indo-Australian and Pacific plates on your cross section (Figure 10.9).

Table 10.6. Earthquakes Associated with the Tonga Trench

Earthquake #	Distance from Trench Axis (km)	Depth of Hypocenter (km)
1	132	66
2	184	83
3	232	116
4	260	150
5	264	216
6	271	250
7	360	300
8	400	350
9	408	383
10	448	448
11	450	500
12	456	516
13	500	550
14	508	600
15	524	624
16	548	667

Figure 10.9. Cross section of latitude 22°S across Tonga Trench (graph not to scale).

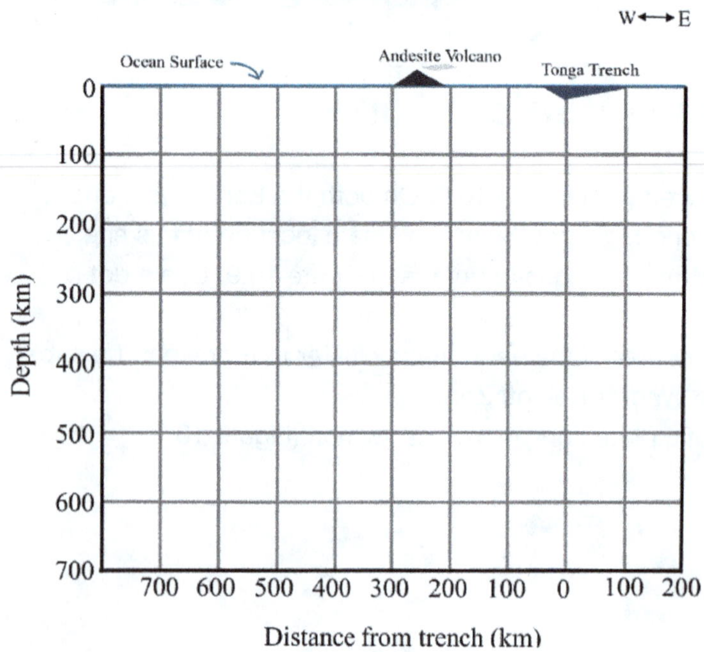

26. Based on the location of the Tonga Trench on both your volcanoes and earthquakes map, what type of convergence is occurring here? Circle one.

   a. Oceanic-oceanic
   b. Continental-continental
   c. Oceanic-continental

27. Predict which plate is subducting (sinking beneath the other). Explain your prediction.

28. Note the Andesite volcano, part of an island arc of volcanoes that parallels the trench axis. Approximately what is the distance, in kilometers, from the axis of the trench to the volcano?

29. Why do volcanoes form at this distance instead of directly above the trench?

10

30. What is the approximate angle of the subduction plane (relative to the surface of the Earth)? Circle one.

   a. 10°
   b. 25°
   c. 45°
   d. 70°
   e. 90°

31. Water released from the subducting oceanic plate partially melts overlying mantle, producing magma that rises to form volcanoes.

a. Using your plotted earthquake data, estimate the depth where melting likely began beneath the volcanic arc.

b. Is this depth considered a shallow, intermediate, or deep hypocenter region?

32. Based on your earthquake plot, on which tectonic plate does the volcanic island arc lie? Explain how your cross section supports your answer.

33. Based on your earthquake plot, which plate descends into the mantle?

34. Compare the earthquake depth patterns of the Tonga and Cascadia subduction zones. Describe one similarity and two differences.

10

35. Using evidence from all four parts of this lab, explain how geologists use earthquakes to better understand Earth's interior and tectonic plate interactions. Your response should reference earthquake location, magnitude, recurrence intervals, plate boundaries, and the Wadati-Benioff Zone. Use evidence for all four parts.

Name: ______________________ Lab Time: ______________ Due: ______________

11 Landslide Hazards

Purpose

Landslides (also known more broadly as mass wasting) are the downhill movement of Earth material. Landslides are a significant natural hazard globally. Each year in the U.S., they cause an average of 25 to 50 deaths and $1-2 billion in damages annually. In the Pacific Northwest, heavy rainfall, steep slopes, seismicity, erosion by rivers, glaciers, and ocean waves, along with a few human factors, have resulted in a significant landslide history.

Objectives

- ❑ Identify physical features that indicate slope instability.
- ❑ Use aerial imagery and LiDAR to identify evidence of past or potential landsliding.
- ❑ Interpret how slope, surface features, and human activity can influence landslide susceptibility.
- ❑ Create and justify a landslide susceptibility map using remote-sensing evidence.
- ❑ Use historical imagery to investigate how landslides can change landscapes, rivers, and human infrastructure.
- ❑ Compare the Oso and Rattlesnake Ridge landslides using evidence from Google Earth.

Key Terms

- ❑ **Remote sensing**: the process of obtaining information about a place from a distance
- ❑ **LiDAR (Light Detection and Ranging)**: a remote sensing method that uses light (in the form of a pulsed laser) to measure distances to the Earth and create precise, 3-dimensional information about the shape of the Earth.
- ❑ **Geologic hazard map**: a type of map that highlights geographic areas that are vulnerable to natural geologic disasters, such as earthquakes, landslides, floods, or volcanic hazards.
- ❑ **Landslide susceptibility**: the relative likelihood that an area will experience landsliding based on characteristics such as slope, geology, drainage, and evidence of previous landslides.
- ❑ **Wadati-Benioff Zone**: the zone of earthquakes that descends into the Earth on a subducting plate.

Materials

- ❑ A device with Google Earth Pro
- ❑ Colored pencils

Part 1. Recognizing Mass Wasting

Viewing Mass Wasting Events

In this section, we will view photos showing mass wasting events or significant mass wasting vulnerabilities. Figure 11.1 offers an internal view of the composition of mass wasting.

Figure 11.1. An internal view of mass wasting.

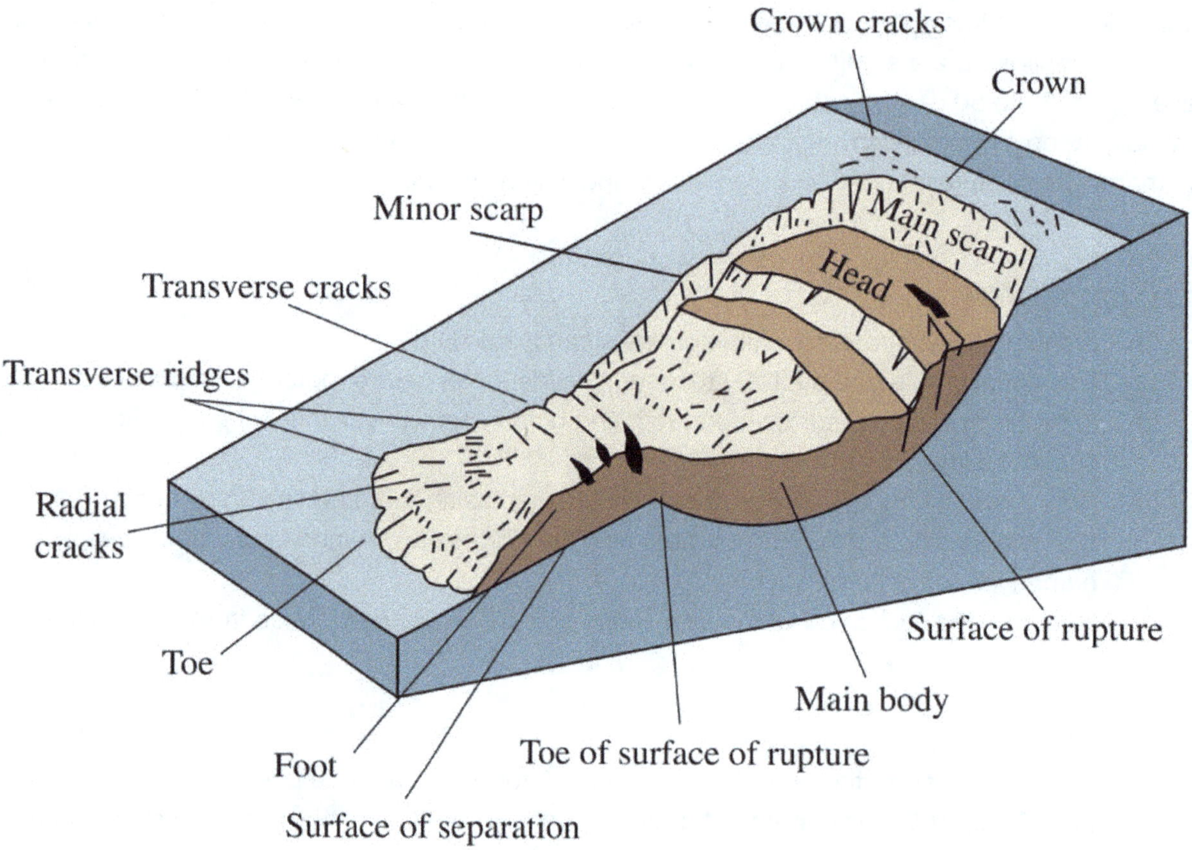

11

Instructions

1. Observe the photos shown and complete the analysis tables associated with each and record your observations in the provided Tables.
    a. Observe: Identify at least 2 features that indicate that the slope may be unstable. Circle or highlight them in the photo. Describe what you see without naming a type of mass wasting.
    b. Interpret: For each feature, explain why it could indicate slope instability.
    c. Predict: What type of mass wasting do you think is occurring or could occur? Explain your reasoning.

Figure 11.2. Slope on the Oregon Coast.

Table 11.1. Analysis Table for Figure 11.2.

Observations	Interpretations	Prediction

Figure 11.3 Slope next to a river.

Table 11.2. Analysis Table for Figure 11.3.

Observations	Interpretations	Prediction

Figure 11.4. Slope on a mountain region.

Table 11.3 . Analysis Table for Figure 11.4.

Observations	Interpretations	Prediction

Part 2. Read the Landscape with Remote Sensing

Examining Landslide Susceptibility in Oregon City

In this section, we will use aerial and LiDAR imagery to examine a portion of Oregon City and evaluate landslide susceptibility in the area.

Instructions

1. Observe and compare the aerial photograph (Figure 11.5) and LiDAR image (Figure 11.6) showing the same area of Oregon City.
2. On Figures 11.5 and 11.6, circle or outline at least 3 features that provide evidence of existing or previous mass wasting.
3. On Figure 11.6, circle or outline at least 2 areas where you think conditions could promote future slope failure.

1. What can you observe in the aerial photograph?

Figure 11.5. Aerial photo of Oregon City, OR near Cascade Highway (SR 213).

2. What can you observe in the LiDAR image?

Figure 11.6. LiDAR image of Oregon City, OR near Cascade Highway (SR 213).

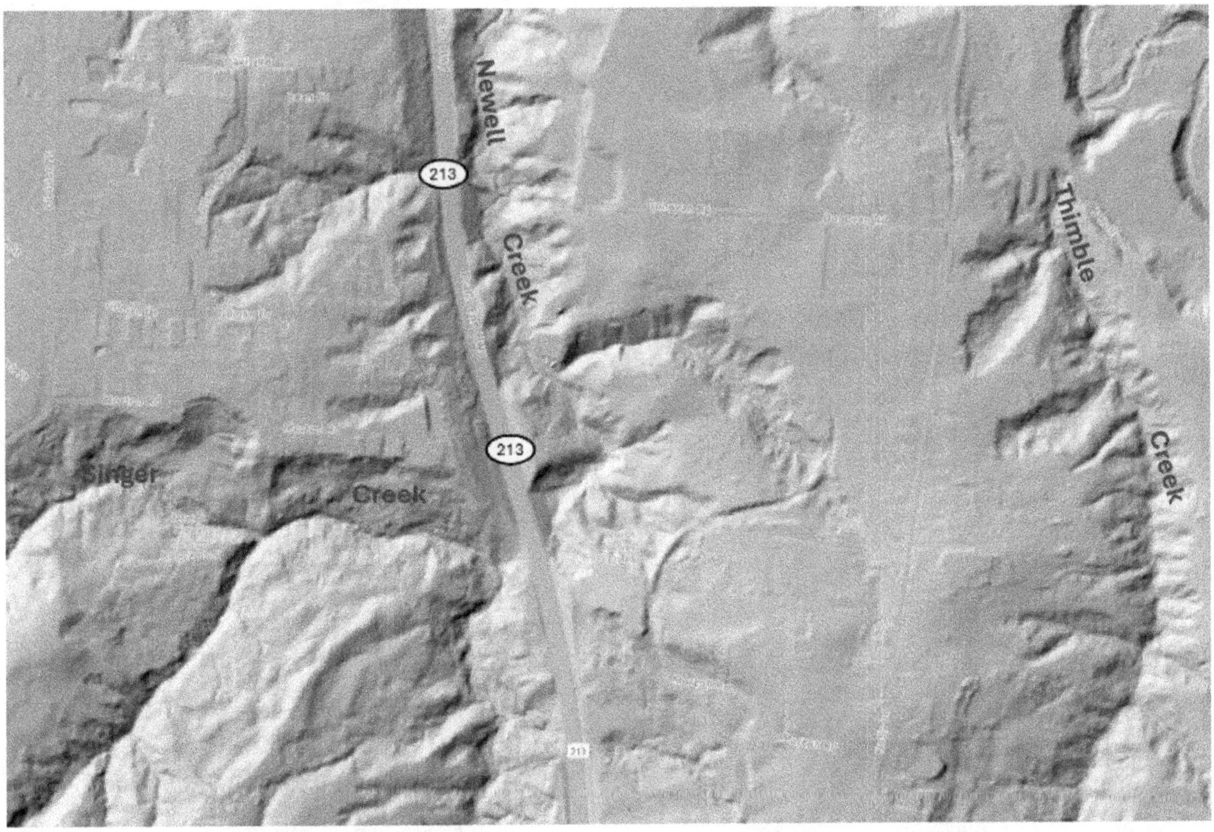

3. Which features are easier to identify in each?

4. What specific features in the landscape provide evidence of existing or previous mass-wasting? Describe at least two.

5. Where do you see conditions that could promote future slope failure?

6. What additional tools, resources, or strategies could you use to identify previous landslides? Describe at least two.

7. Based on the conditions you observed in this area, what human interventions could reduce the likelihood or consequences of mass-wasting? Describe at least two and explain how each could help.

Part 3. Landslide Susceptibility Mapping

Classifying Landscapes

Here, we will use the evidence you gathered in 11.2 to classify the landscape by its relative susceptibility to landsliding.

Instructions

1. On Figure 11.7, use colored pencils to shade the map according to the color scheme below. This map represents your interpretation of landslide susceptibility based on the available evidence. It is not a prediction of exactly where or when a landslide will occur. Your entire map should be colored.

- ❑ Red: areas with evidence of existing or previous landsliding
- ❑ Orange: areas without obvious evidence of previous landsliding, but with a high susceptibility to future landsliding
- ❑ Yellow: areas of moderate susceptibility
- ❑ Blue: areas of low susceptibility based on the available evidence

2. Add the following elements to your map:

- ❑ Name/title (above it)
- ❑ Key/color code (below)
- ❑ North arrow (in a corner of your choice)

8. Identify one area you classified as having high susceptibility and explain what evidence led you to that classification.

9. Identify one area where you were uncertain about the susceptibility classification. What additional information would help you make a more confident assessment?

Figure 11.7. LiDAR image of Oregon City, OR near Cascade Highway (SR 213).

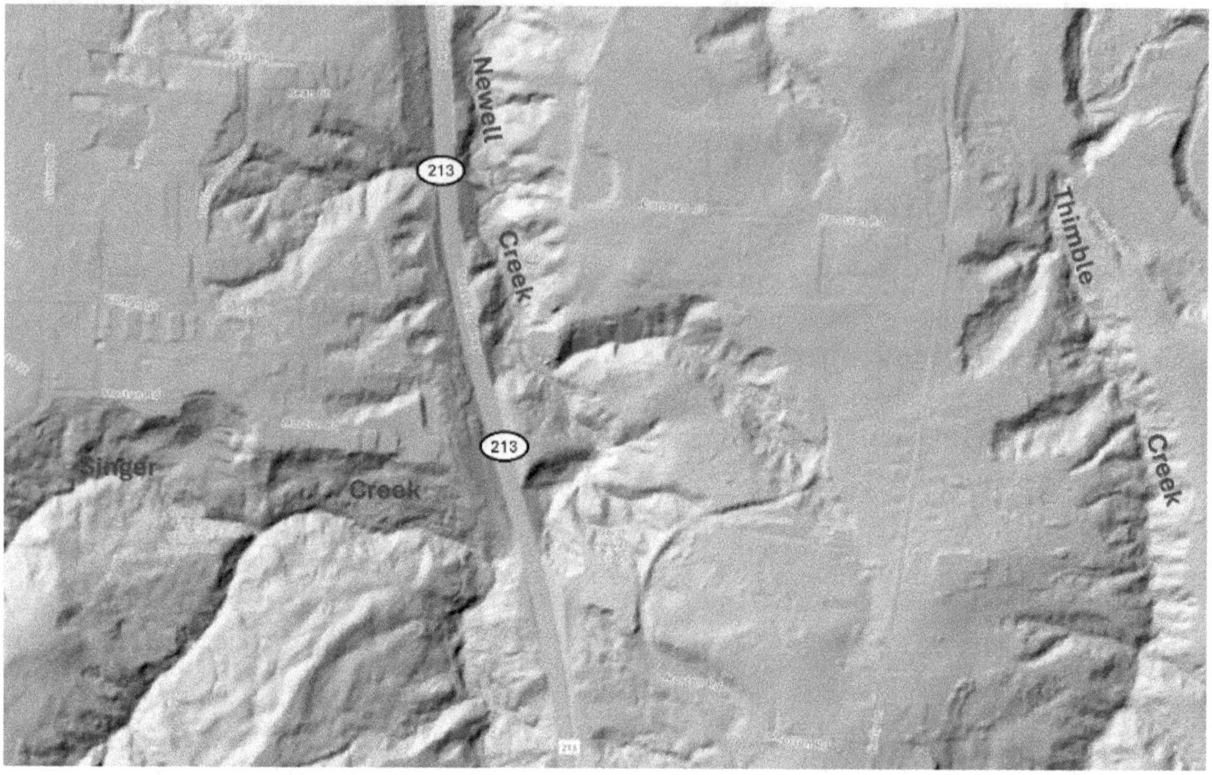

Part 4. Pacific Northwest Landslides

Introduction

In this section, we will investigate two recent Pacific Northwest landslide events using Google Earth. Places where you need to provide an image are highlighted in yellow.

11

Oso Landslide, Washington

The significant and tragic 2014 Oso landslide event dramatically altered both the hillslope and the Stillaguamish River. In this investigation, you will use elevation data and historical imagery to reconstruct how the landslide changed the landscape and affected the Stillaguamish River and nearby community.

Instructions

1. As in the previous Google Earth labs, be sure to check that your settings are correct (see Lab 7).
2. Center your map on the state of Washington and observe the state's general layout. Note the location of the Cascades, Puget Sound, and the wet vs dry areas.

3. Using the search bar, navigate to the "SR 530 Oso Slide Memorial". Zoom out and to the north to find the slide itself.
4. Spend some time observing this area to get a sense of what happened. Adjust the perspective from map view to oblique and back view a few times.
5. Visually identify the area of erosion, area of deposition, and the river that cuts through it (almost entirely on the north side of State Road 530).
6. Using the "Path" tool, make an elevation profile[1] that runs down the axis of the landslide from the crown to the toe (use at least 4 control points since it is not perfectly straight). Observe how the slope and elevation change along your profile line (move your cursor along the profile).
7. Take a screenshot showing the full length of your profile line on the map and the elevation profile and insert it into the answer sheet.
8. Find the following landslide features. Use the "Placemark" tool and create a new pin. Name each pin as the feature it indicates.
    - » crown
    - » head
    - » head scarp and minor scarp
    - » toe
    - » transverse cracks[2]

9. Take a screenshot with these labeled pins and insert it in the answer sheet.
10. Using the "Historical Imagery" tool, go back in time and view how the hillside, surrounding landscape, and stream channel have changed from the 1980s until the most recent image. Spend some time observing the changes over the decades.
11. Zoom to an eye alt of about 8000 ft[3] and observe the area just northwest of the slide. Note how vegetation cover changed from the 1980s until the early aughts (00s).

10. Describe the specific changes in vegetation you see and consider how they could impact slope stability.

11

---

1. Revisit the Intro to Google Earth lab for an explanation of how to make an elevation pro

2. These become more obscure over time as vegetation reestablishes and they smooth out due to erosion and deposition. However, they can still be seen as linear features with less vegetation that sometimes fill in with water. They generally form perpendicular or roughly perpendicular to the direction of landslide movement.

3. Use the historical imagery slider bar to move through time, but at this high altitude, it is important to check the date of the image by hovering your cursor over an area and checking the information bar at the bottom right of your screen.

12. Go to an image in 2005. Using the "Path" tool, carefully trace the river in this vicinity for about 2 miles (with the Oso landslide in the approximate center). Name your path "Stillaguamish River 2005," change its color to blue, and its width to 2.0.
13. Advance to an image from 2006. Observe the significant changes in the hillside and the river's location.
14. Make another path of the river with the same starting and stopping points. Name this path "Stillaguamish River 2006," change its color to yellow, and its width to 2.0.
15. Close the "Historical Imagery" tool.
16. Take a screenshot showing the 2005 and 2006 locations of the river superimposed on the current location. Insert it in the answer sheet.

11. Observe how the hillside and Stillaguamish River changed over this approximately 1-year period. Describe these changes.

17. Go to the July 2013 image and find the small community located south of the river and mostly north of SR 530.

12. Using the "Historical Images" tool, look at the few years before 2013 and determine when they were built. Give the year or a specific range.

11

18. Go to the April 2014 image and note that it was taken only a few weeks after the landslide occurred.

13. Compare this image to the July 2013 image and describe, in detail, the land changes you see.

14. **Using the** same two images, describe how the river changed—flow, location, dimensions, and appearance of the water upstream and downstream of the slide.

19. Using the "Path" tool, measure the farthest extent of the landslide—from the crown to the toe[1].

15. Give this distance in miles.

20. In the April 2014 image, use the "Path" tool and circle the zone of deposition of the landslide.
21. Go back to the July 2013 image—count the number of homes/buildings that appear to have been buried in the debris.

16. Give the approximate number of structures.

Rattlesnake Ridge Landslide, Washington

Here we will look at another landslide event in Washington State using Google Earth Pro.

> Note: Carefully read the directives below and follow the prompts.

Instructions

1. Open Google Earth Pro.
2. Using the search bar, navigate to the following coordinates: 46°31'29" N, 120°28'00" W. This will take you to Rattlesnake Ridge in the Rattlesnake Hills near Yakima.
3. Spend some time observing this area to get a sense of what happened. Adjust the perspective from map view to oblique and back view a few times.

1. Follow the flow by using multiple control points and watch the distance increase in the "Measurements" tab.

17. Describe the appearance of the land in the immediate area.

4. Zoom in to an eye alt of 5000-7000 ft, with the slide area at the center of your screen. An oblique (tilted) view will be useful.
5. Using the "Historical Imagery" tool, go back in time and observe the changes near the top of the ridge.

18. Describe the changes in the appearance of the land between 2017 and 2018. Be specific.

19. Describe two anthropogenic (human-caused) changes that are apparent between 2017 and 2018. Hint: these changes were in response to the event. Be specific.

6. Adjust your view to look obliquely at the west side of the ridge and zoom to an eye alt of about 1370 ft.

20. Using the "Historical Imagery" tool, describe a change that occurred to the hillside and give the 1-2 year date range during which this change occurred.

21. How are the causes, characteristics, and consequences of the Oso and Rattlesnake Ridge landslides similar? How are they different?

22. What evidence can geologists use to identify areas where landslides may pose a hazard to people and infrastructure? Use at least two specific examples from this lab.

11

Name: ______________________________ Lab Time: ________________ Due: ________________

12 Landslide Hazards Part II

Purpose

To apply our knowledge of mass wasting and to assess the dangers in areas of the Pacific Northwest prone to mass wasting events.

Materials

- ❑ Google Earth
- ❑ Provided images

Part 1. Rattlesnake Ridge Landslide

Answer the questions using the provided materials.

Instructions

1. Using Google Earth, go to Yakima/Union Gap, WA. South of Union Gap, along the Yakima River, there is a landslide (called the Rattlesnake Ridge Landslide). The slide is located along a ridge east of the river, with a quarry at its toe.

1. Looking at the ridge and the slide, describe what you see.

2. Figure 12.1 is a cross section of the rocks in the area. What type of mass wasting events are likely occurring here? Explain your answer.

Figure 12.1. A cross section displaying the Rattlesnake Ridge Landslide.

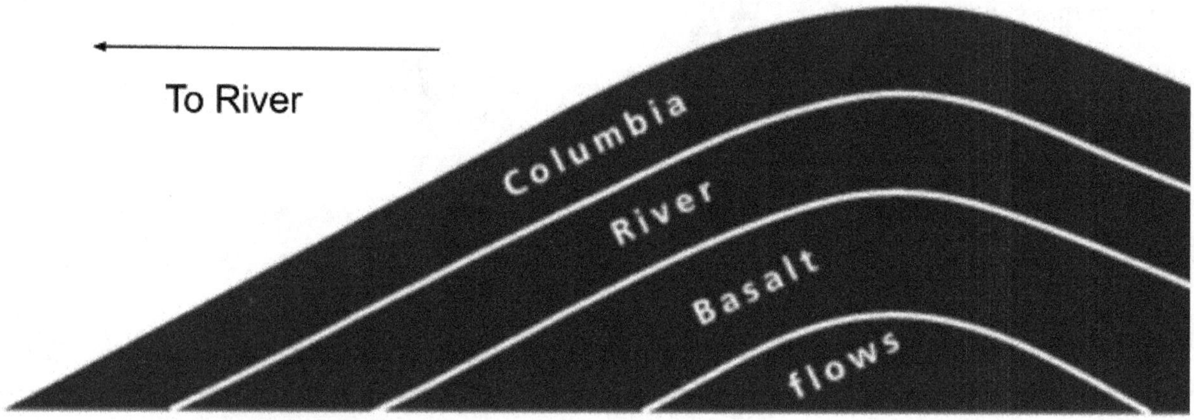

3. Look along the ridge beyond the obvious rupture point above the quarry. What mass wasting concerns do you see as you follow the ridge to the north?

4. Describe the river nearby. Anything strange you notice about it?

12

Part 2. Using SLIDO (Statewide Landslide Information Layer for Oregon)

Instructions

Go to https://gis.dogami.oregon.gov/maps/slido/ or use QR Code 12.1.

QR Code 12.1. Oregon Department of Geology and Mineral Industries SLIDO

1. On the layers option on the right, turn off "Landslide Inventory."
2. Expand the "Landslide Susceptibility" menu.
3. Turn off all sub-options except "Regional Landslide Susceptibility." This is a statewide assessment of landslide concern.
4. Make sure the main "Landslide Susceptibility" option stays selected.

5. Look along the coast for areas of high susceptibility. Select one area and list it here.

5. Using Google Earth and the "Bare Earth LiDAR Hillshade," explore this area you just found.

6. Explain what you can see that suggests this area has a high potential for landslides.

7. Given the area, what other issues may be influencing the landslide concern for this location that would not be found/visible on your maps?

6. Zoom out a bit so you can see the whole state.
7. Deselect "Regional Landslide Reference."
8. Select "Detailed Susceptibility Reference Maps" and "Susceptibility of Shallow Landslides."
9. Select one of the areas outlined in black. As you zoom in, you should find more detailed hazard maps of that outlined area.
10. Go back and turn on "landslide Inventory," open the menu, and deselect "deposits." Leave "scarps" and "head scarps" on.

8. Find areas with a scarp but lower landslide susceptibility. Provide an explanation.

Note: The "deposits" option, as well as the previously used "LiDAR" option, may be useful here.

9. Find some areas where there is a scarp and a high susceptibility of landslides. In the context of your previous answer, provide an explanation.

Note: The "deposits" option, as well as the previously used "LiDAR" option, may be useful here.

Part 3. Washington State Geology Portal

Instructions

Go to the following link: WA Interactive map or use QR Code 12.2.

QR Code 12.2. WA State Interactive Geology Map

1. Deselect the “Surface Geology” option if it is selected.
2. Select the “Landslide” option. It should automatically turn on the WGS-Protocol subheading. If anything else turns on, turn it off.
3. From a statewide perspective, you should see some outlined areas (mainly the west side, some along the Columbia). Zoom in and look at what is present.

Note: You should see an option to turn on the legend along the menu at the bottom.

10. Several different landslides are marked (not the dots for recent landslides). Describe what they are and where you tend to find them.

Figure 12.2. Screenshot of the Washington State interactive map.

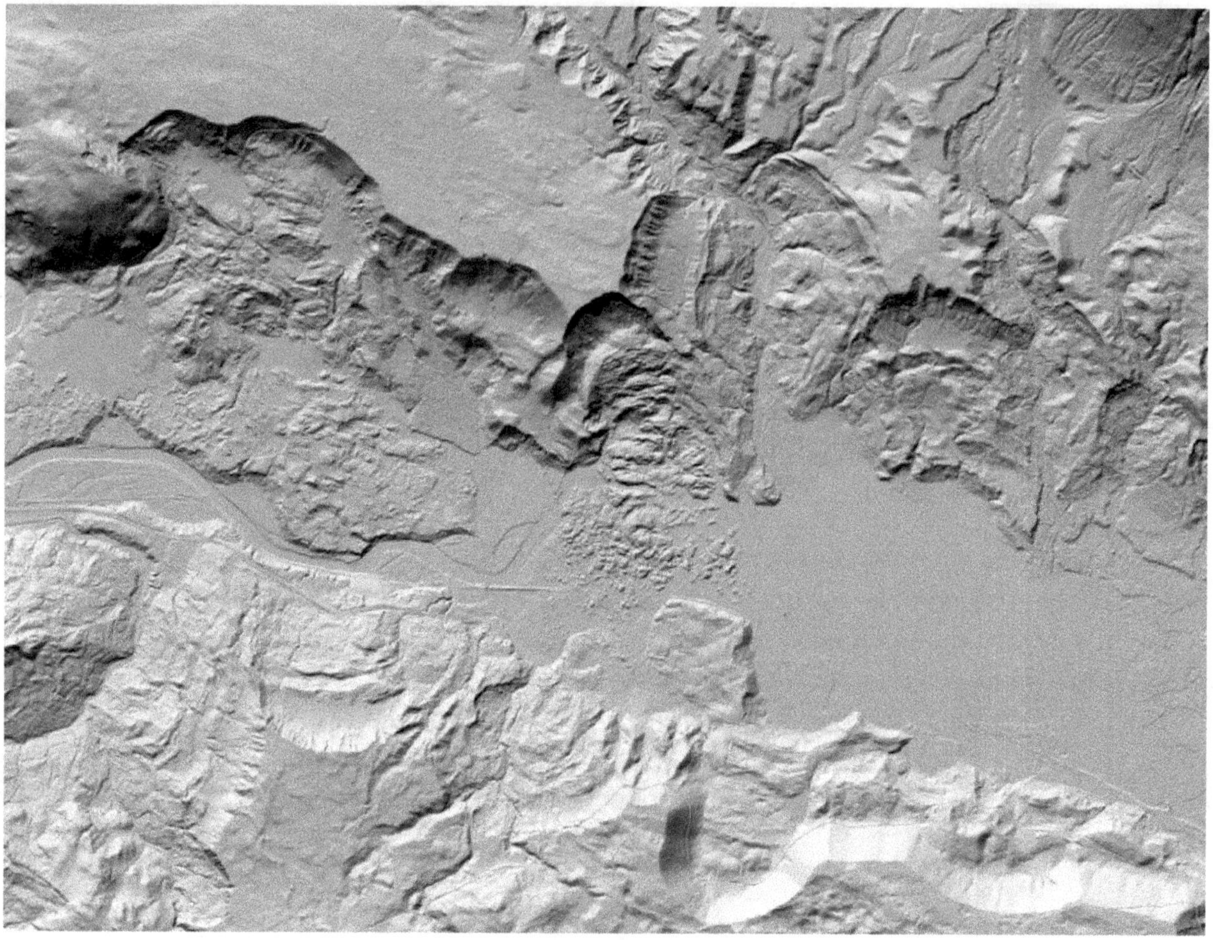

4. Look at the various shapes of the hills to the North and South of Figure 12.2, and the features of the valley between.

11. Now that we've talked about this a little bit, describe the hills. What prominent feature seems to be present along the elevated section regularly?

12. What do you see in the valley? Are there any patterns in the valley that seem to be connected features?

13. What are these features we're looking at? Explain your answer.

5. Go to our WA interactive map. Work your way to this area (by Rowan, WA, if you're able to see it on the map). Look through the landslide information provided.

14. What similarities and differences do you see between our original LiDAR image and this interactive map?

6. Look closely at the valley in the LiDAR image; you should see the line representing the river running through the area.

15. Why might this add to the concern and hazard we're looking at above?

Here's some fun. We're looking at the area around the Oso Landslide in WA in 2014. In our LiDAR image, the landslide is in the middle of the picture.

7. Go to Google Earth and look for the location of the failure (it'll be fairly obvious).
8. On the top slide, you should see the option for historical imagery. Go back to an image before 2014.

16. Explain some differences and similarities between the images.

Let's see if we can spot the discrepancy.

9. Look at the images between Google, our LiDAR image (which is the base map in our WA interactive map), and the landslide deposits.

17. There is something a bit odd about the river and its representation. What's the issue?

12

Name: ______________________ Lab Time: ______________ Due: ______________

13 Volcano and Earthquake Hazards

Purpose

To apply our knowledge of earthquakes and volcanic eruptions to assess the dangers in areas of the Pacific Northwest.

Materials

- ❑ Maps, all online

Part 1. Volcanic Hazards of the Pacific Northwest

Instructions

Go to the following website: https://www.usgs.gov/natural-hazards/volcano-hazards/ or use QR Code 13.1.

QR Code 13.1. USGS Northwest Volcano Hazards Maps

1. Navigate to the page of a Pacific Northwest volcano that is associated with eruptions along the Cascadia Subduction Zone. Do not use Mount Saint Helens or Crater Lake.
2. Once on the page for the mountain, go to the "Science" link in the left menu.
3. Within the subheading, you should find a "Hazards" option.
4. There should now be a "Hazards Summary" thumbnail linked to a page that includes a hazard map for the volcano. If it doesn't, you'll need to keep exploring or find another volcano.
5. Answer the following questions.

1. What sort of information is included in the hazard map provided? Include which volcano you're looking at.

2. How detailed is the map? Do you think it would be useful in hazard planning at the state level? How about the country or city level? Why or why not?

3. Summarize the volcanic hazards of the area (based on the map, not looking them up). Which ones are a problem for the area? Do they cover a narrow area? Broad area?

4. What features are associated with the lahar hazards on your map?

5. Look up some of the cities and towns on the map that are within or near the hazard areas. List their populations here (at least three).

6. Under the "Science" link, click "Volcano Monitoring."

6. What sorts of things are they doing to keep track of your volcano?

Note: If they don't have a specific monitoring section, explore the links and see what they do have that discusses volcano activity.

7. Click through the various monitoring links or any available information.

7. Has any activity been monitored around your volcano in the last 50 years? What sort? How was it discovered/monitored?

8. Click through the various monitoring links or any available information.

8. Has any activity been monitored around your volcano in the last 50 years? What sort? How was it discovered/monitored?

9. Based on the map and the monitoring information you have available, how concerned would you be about an increase in activity levels? This refers to seismic activity, slope changes, or gas emissions—those sorts of things. At what point would you consider warning nearby towns? Evacuating?

10. What preventative measures would you put in place for the local communities to reduce the potential impact of a future eruption? Not to make the eruption smaller, but to reduce the damage and/or casualties in the case of an event. Explain your answer.

11. Select another Cascade volcano. Compare and contrast the hazards between the two. How are they similar, how are they alike? Based on their hazard maps and monitoring data, which would you think is the more concerning volcano in terms of eruption potential? In terms of risk (damage)? Explain your answer.

13

Part 2. Earthquake Hazards

Instructions

Go to the following website: https://earthquake.usgs.gov/nshmp/source/model-maps or use QR code 13.2.

QR Code 13.2. USGS Model Maps

1. Spend some time playing with the model's features. Pay particular attention to the option to turn the faults and hazard maps on and off. In the end, you'll want a map that looks like Figure 13.1 (with the faults visible).

Figure 13.1. USGS frequency of damaging earthquakes across the U.S.

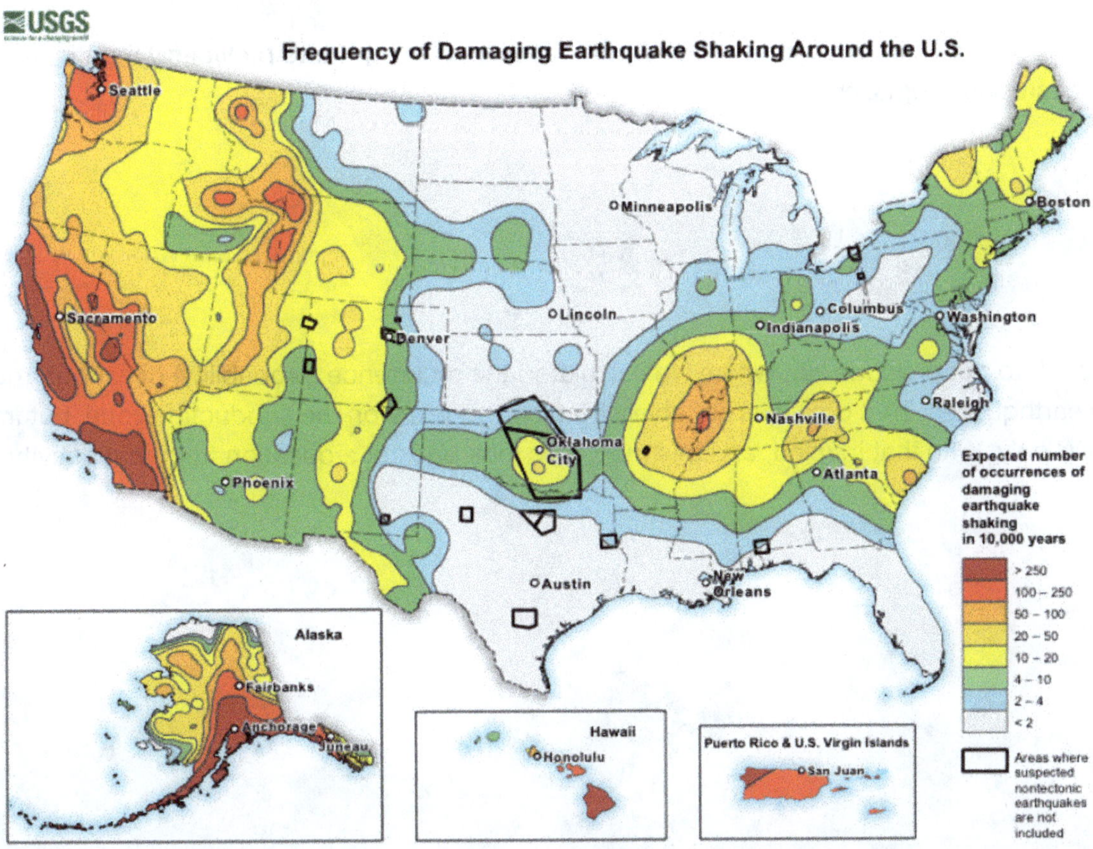

2. Answer the following questions.

12. Looking at the Pacific Northwest (Oregon, Washington, Northern California), what areas seem to be at higher risk for earthquakes? What patterns do you see?

13. While we expect a hazard associated with our local subduction zone, what areas of high hazard seem not to be related to it? Hint: This is a map of shaking, so consider: if nothing else were occurring, what would we expect the pattern to be relative to our offshore plate boundary?

14. How do the various faults in Oregon and Washington compare to previously mentioned hazardous anomalies?

Think back to our previous lab, where we calculated the recurrence (probability) of various-magnitude earthquakes. That data was not solely from earthquakes on the subduction zone, but from the PNW only. With that in mind, and this map, let's look back at a question from that activity.

15. Given what you've found, should public officials, emergency planners, engineers, and individuals focus on earthquake preparedness efforts for M 8-9 earthquakes? How about M 6-7 earthquakes? Explain your reasoning in detail. Please write at least one paragraph.

Part 3. Tsunamis

Instructions

Go to the following website: https://nvs.nanoos.org/TsunamiEvac or use QR Code 13.3.

QR Code 13.3. Tsunami Evacuation Zone Map

1. Spend some time playing with the model's features. Click on some of the rectangles (which zoom in to specific areas) and get an idea of the various tsunami concerns for areas along the PNW coast.

Go to https://www.ncei.noaa.gov/maps/ttt_coastal_locations/ or use QR Code 13.4.

QR Code 13.4. NCEI Tsunami Travel Times to Coastal Locations Viewer

2. Play around with the map option and select different locations for the origin of a hypothetical earthquake/tsunami (stay within the Pacific Ocean region).
3. Be sure to select some options in Alaska, Seaside, OR, and Crescent City, CA.
4. Take a look at some origins on the other side of the Pacific as well.
5. Answer the following questions.

16. Assuming a significant earthquake along the Pacific Northwest plate boundary, how long, according to our maps linked above, before the follow-up tsunami hits the coast (assuming one is generated)?

17. Should we be concerned about potential tsunamis originating elsewhere in the Pacific affecting the PNW? What location would be most concerning? Explain your answer.

18. Describe the tsunami hazard zones in the PNW. Where are they? Do they cover a broad area or a narrow area? Why?

13

19. What factor controls the hazard zones between tsunamis that originated across the Pacific vs. tsunamis generated along the PNW subduction zone? Yes, some originated further away, but focus on the zones themselves. How is one different from the other?

20. In areas where the hazard zones extend farther inland, what allows this? Why might the locations of these inland hazards be a major concern for the PNW coast?

21. Based on the information available to you, how concerned would you be about tsunami hazards in the following areas? Would these areas need significant evacuation routes? Should education or awareness plans be in place? Should they include tsunami hazards as a factor in their building codes and requirements? Explain your answers.

- ❑ Portland, OR

- ❑ Salem, OR

13

- ❑ Seattle, WA

- ❑ Tacoma, WA

- ❑ Astoria, OR

- ❑ Newport, OR

- ❑ Seaside, OR

13

22. The Puget Sound is significantly inland. Why is there such significant concern about local tsunamis in the area? Look back at the Earthquake faults and hazards map.

Part 4. Foreshadowing of a Future Topic

Instructions

1. Look at the LIDAR map of an area in western WA in Figure 13.2.

Figure 13.2. A LiDAR map of western Washington state.

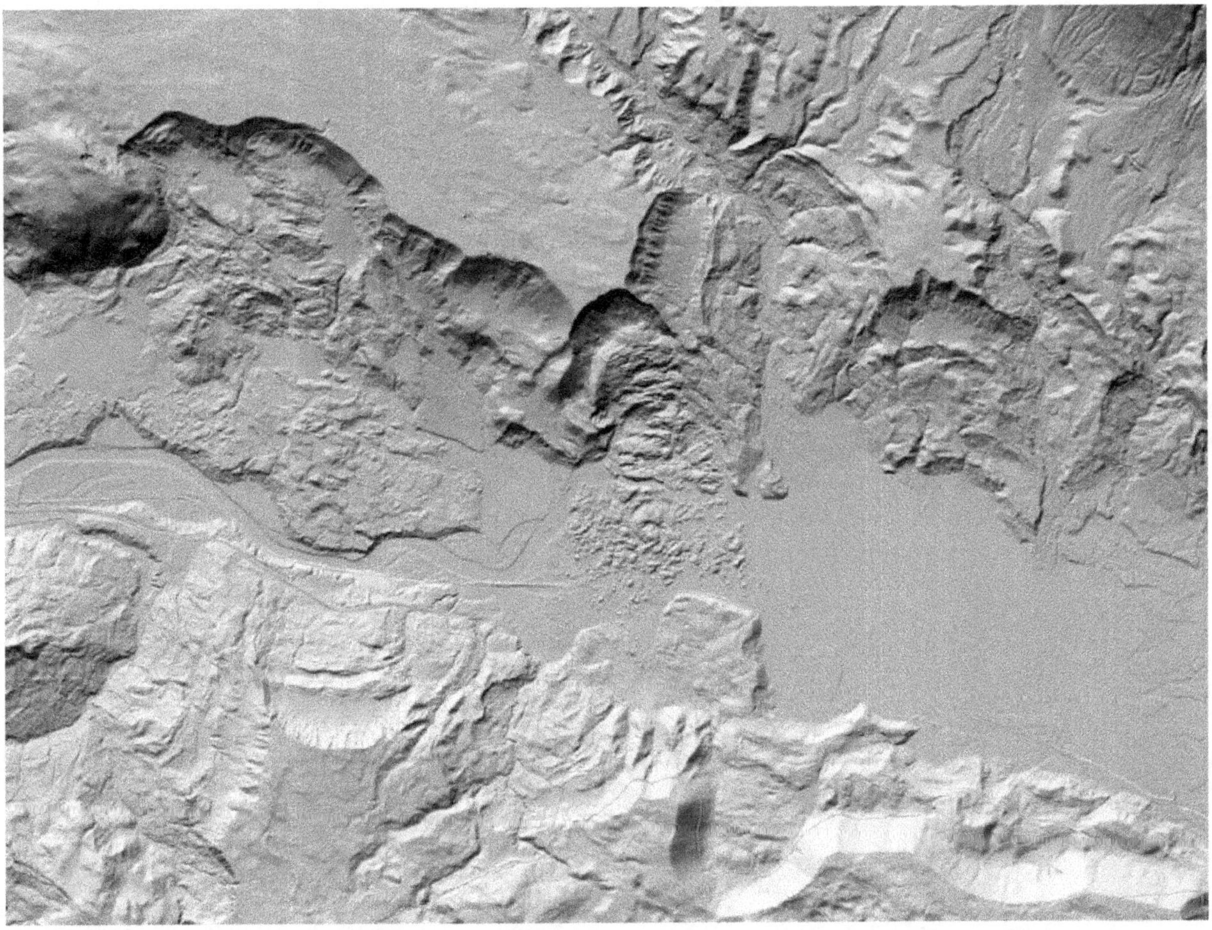

2. Figure 13.2 uses technology to see the surface terrain without vegetation, which obscures many features in the western part of the PNW.
3. Look at the various shapes of the hills to the North and South of the image, and the features of the valley between.

We haven't quite done this topic yet…but that's okay.

4. Look at the image and take a stab at it. You're not going to be graded on whether you're right or wrong, but on the effort you put into understanding and explaining the image.
5. Answer the following questions.

23. Describe the hills. What prominent feature appears regularly along the elevated section?

24. What do you see in the valley? Are there any patterns in the valley that seem to be connected to previously mentioned features?

25. What are these features we're looking at? Why are there so many of them? Why do you think this?

26. Look at the valley closely. You should see the line representing the river through the area. Why might this add to the concern and hazard we're looking at above?

13

Name: ______________________________ Lab Time: __________________ Due: ____________________

14 Field Trip: Oregon Coast

Stop 1: Cascade Head

We are not going to hike to the top of this head, but if you get the chance, it's well worth it. However, from where we are parked, look out slightly north.

1. Draw the rock outcrop below.

2. What type of rock do you think this is? Is it related to the Columbia River Basalts?

3. From this new vantage point, describe and sketch the river we are looking out on.

4. Is its gradient steep or flat?

5. Is its velocity fast or slow?

6. Relate the river's behavior back to our coastal tectonics and indicate whether the river is eroding vertically or laterally. How can its characteristics be changed?

Stop 2: Proposal Rock, Neskowin

Get as close as you can to Proposal Rock.

7. Describe the texture of the rock that you see.

8. How is this rock different from other basaltic flows you've seen?

9. What would have to happen to a basalt flow to create the rocks that you are seeing?

Stop 3: Cape Kiwanda, Pacific City

10. Compare the color of the cape to Haystack Rock in the ocean. Based on your observations, are they the same rock?

11. What type of rock are the caps and Haystack Rock, respectively? Are they igneous, sedimentary, or metamorphic?

12. How exactly do you think the sand dune formed behind the cape?

14

Stop 4: Nestucca Valley High School, Cloverdale

Notice the topography of the surrounding valley.

13. Describe it, making sure to mention the valley walls and the outlet for the Nestucca River.

14. If there were a large magnitude earthquake and resulting tsunami, how would this valley's configuration affect the outward flow of tsunami waters?

Stop 5: Mt, Hebo, Hebo (Possible Stop)

Notice how flat Mt. Hebo is. The highest peaks in the Oregon Coastal mountains have similar caps, allowing them to maintain their height, like Mary's Peak near Corvallis.

15. What do you think has protected this particular peak from erosion?

14

Name: ______________________ Lab Time: ______________ Due: ______________

15 Field Trip: Mt. St. Helens Report

Knowledge Review

1. Define viscosity and explain what influences the viscosity of magma.

2. List the 3 aphanitic volcanic (extrusive) rocks we studied. Put them in order of increasing silica content.

3. What is Dacite? Hint: it's not in your book—review your notes or ask an instructor after you have answered the second question.

15

Driving North (I-5)

4. Are any volcanoes visible from the commute? If so, name all that you view below.

5. We cross into Washington state on the I-5 interstate bridge. What river are we crossing?

6. What direction is the river flowing?

Driving East (State Route 504)

7. What types of rocks or lava flows do you see en route that we have identified in class?

8. What is the name of the national forest we are entering as we approach the Johnston Ridge Observatory?

Johnston Ridge Observatory

9. Who is the namesake of this observatory? Explain.

1. Find the "Topographic model with colored lights."

10. Listen to the description of the eruption and fill in Table 15.1. Indicate the colors of lights used and describe each event and how it changed the landscape.

Table 15.1. Description of Mt. St. Helens' Eruptions

List the Eruptive Events in Order	Color of Eruptive Event	Describe Event. How did it change MSH or the landscape?
1		
2		
3		
4		
5		
6		

2. Find the exhibit "There are Different Recipes for Eruption." Read the panels and examine the rocks.

11. Circle the volcano that explodes more violently:

Kilauea OR Mt. St. Helens

12. Which type of lava contains the least amount of silica?

Basalt OR Dacite

3. Find the panel "Lava Domes Rebuild Mt. St. Helens."

13. Circle which type of dacite lava is destructive:

Gas-rich dacite lava OR Gas-poor dacite lava

14. Describe how the formation of the 1980–1986 and 2004–2008 lava domes were similar and different.

4. Watch the video entitled "Crater Glacier."

15. What feature created by the destructive May 18, 1980 eruption allowed the glacier to form?

16. What do you think will happen to the Crater Glacier in the future—will it retreat or advance? Why?

5. Go to the "Shattered Tree Stump" and find the display about tree rings.

17. What do the tree rings tell us about MSH's eruptive history?

15

6. Find the "New Answers to Old Mysteries" exhibit.

18. What two eruptive events at MSH led to the discovery of similar events at other volcanoes?

19. How is MSH rebuilding itself?

Eruption Trail

Stop 1

1. Find a stump or log beside the trail between the trailhead and the first interpretive sign. Answer these questions.

20. Towering evergreen trees grew here before the eruption. The lateral blast created 3 distinct "impact zones." Make observations to determine the impact zone here. Circle one.

- **a.** Tree Removal Zone: Trees uprooted, shattered, and swept away from 0-7.5 miles away from the volcano.
- **b.** Blowdown Zone: Trees are blown down in distinct patterns. Ridges and topography direct the blast 2.5 to 15.5 miles away from the volcano.
- **c.** Scorch Zone: Hot gases scorch trees, creating a standing dead forest, 2.5 to 17 miles.

Stop 2

1. Find a sundial-like landscape locator is at the circular plaza on the top of the hillside.
2. Find Coldwater Peak on the landscape locator and observe the blast features below and around it.

21. Circle which "impact zone" Coldwater Creek belongs in.

- **a.** Tree Removal Zone
- **b.** Blowdown Zone
- **c.** Scorch Zone

15

22. Explain how a blown-down tree is like a compass needle.

3. Look around for blown-down trees on the nearby ridges. Use the locator map to identify what direction the blown trees lie (point towards).
4. Use the locator map and find east. Look east down the length of Johnston Ridge and find the mounds on top of the ridge.

23. What are these called and why are they there?

Hummocks Trail

Background Information

The May 18, 1980 eruption began when a huge landslide fell from the bulging north side of Mount St. Helens. Three enormous slabs of rock, called slide blocks, fell downward in quick succession.

Slide Block I is dominated by dark gray andesite lava rocks and black or red basalt lava rocks from the volcano's north flank. It also contains minor amounts of light gray dacite rocks from the summit.

Slide Block II is dominated by light gray dacite rocks from the former summit.

Moments after Slide Block I came to rest, Slide Block II pushed it out of the way like a gigantic snow plow. Due to this violent collision, Slide Block I is largely found along the sides of the valley, while Slide Block II dominates the center of the valley.

Slide Block III is dominated by tan dacite lava rocks from the interior of the volcano. This slide block was violently shoved outward by a lateral blast, pushing the landslide 13.5 miles down valley (4 miles past Slide Blocks I and II). Few intact hummocks from this slide block are found west of the hummocks trail—they were broken apart and mixed together during their turbulent journey.

Stop 1

1. After reading the background information, look at the trailside rocks by the bench.

24. Based on the color of rocks in your surroundings, from which slide block did these rocks originate? Circle one.

a. Slide Block I

b. Slide Block II

c. Slide Block III

Stop 2

Johnston Ridge should be in sight, and MSH is straight ahead. The rocks surrounding the trail came from the same slide block as above.

25. The landslide began on the north side of the volcano, but traveled only 5.5 miles north. However, it traveled 13 miles westward. Why do you think this happened?

Stop 3

Notice the hummock that is half-light tan and half blackish-red. This is the two-toned hummock.

26. Based on the color, name the types of rocks that might be found on each side of this hummock.

Acknowledgments

Image Acknowledgments

Unless otherwise stated, figures created and copyrighted by the authors or Chemeketa Press. © 2026.

Lab 1

Figure 1.1. "Ice building in Tualatin on Sunday" by Ron Schaaf via Chime In. https://katu.com/news/local/2021-ice-storm-took-out-nearly-18-of-salem-tree-canopy

Figure 1.4a.+ b. "Above Crater Lake (cropped)" by Athleticamps. Wikimedia Commons. https://commons.wikimedia.org/w/index.php?curid=74454500.

Figure 1.5. "Fumaroles in Porcelain Basin" by NPS / Jime Peaco. https://www.nps.gov/places/000/fumaroles-in-porcelain-basin.htm?utm_source=place&utm_medium=website&utm_campaign=experience_more&utm_content=large

Figure 1.6b. "Crater Lake" by Annie Scott / USGS. Public Domain. https://www.usgs.gov/media/images/crater-lake-6?utm

Figure 1.8. "Mt St Helens, Washington State. The photo on the left (older) and the one on the right (younger) were taken just a few months apart" by USGS.

Figure 1.9. "Dirt road in Mojave Desert near Ridgecrest, CA" by USGS.

Figure 1.10. "Chugach National Forest: Ghost Forest" by Sabina Ayne. https://aquamarineeyes.wordpress.com/2015/05/08/force-of-nature-ghost-forest/

Figure 1.11. "Liquefaction at Niigata" by Ungtss. Wikimedia Commons. https://commons.wikimedia.org/w/index.php?curid=18730337

Figures 1.12-1.13. Google Earth Pro.

Lab 7

Figures 7.1-7.4. Google Earth Pro.

Figures 7.6-7.7. Google Earth Pro.

Lab 8

Figures 8.1-8.4. Google Earth Pro.

Figure 8.5. "Cascade Range" by USGS. https://www.usgs.gov/media/images/cascade-range

Figure 8.6. "Floods of Lava" by USGS / Barry. https://www.nps.gov/laro/learn/nature/geology.htm

Lab 10

Figure 10.6. "Distribution of earthquake epicenters (M 4.0 and greater from 1990-1996)" by Rice University. https://www.plateboundary.rice.edu/downloads.html.

Figure 10.7. "Distribution of volcanic activity" by Rice University. https://www.plateboundary.rice.edu/downloads.html.

Figure 10.8. "Satellite images of the Tonga Trench area. 9a. Regional view of the South Pacific area. The red box highlights the map area in 9b. 9b. Tonga Trench near the Kingdom of Tonga. Note the 22°S latitude line. The red triangle represents the andesite volcano referenced in this activity," by NOAA.

Lab 11

Figure 11.1. "Landslide Types and Processes," by USGS. https://pubs.usgs.gov/fs/2004/3072/fs-2004-3072.html.

Figure 11.5. "Aerial photo of Oregon City, OR near Cascade Highway (SR 213)" by Google Earth Pro.

Figure 11.6. "LiDAR image of Oregon City, OR near Cascade Highway (SR 213)" by Oregon Department of Geology and Mineral Industries. LiDAR Data Viewer. https://geohub.oregon.gov/apps/oregon-geo::lidar-data-viewer/explore

Lab 12

Figure 12.2. "Landslide LiDAR Northwest Washington," by USGS. https://www.usgs.gov/media/images/landslide-lidar-northwest-washington

Lab 13

Figure 13.1. "Map showing the predicted frequency of damaging earthquake shaking," by USGS. https://www.nps.gov/articles/000/lehman-caves-little-understood-but-world-class-part-2.htm

www.ingramcontent.com/pod-product-compliance
Lightning Source LLC
Chambersburg PA
CBHW081202150926
51527CB00001B/1
9781955499439